The Baculovirus Expression System

The Baculovirus Expression System

A laboratory guide

L. A. King
School of Biological and Molecular Sciences, Oxford Polytechnic, Oxford, UK

R. D. Possee
Natural Environment Research Council, Institute of Virology and Environmental Microbiology, Oxford, UK

SPRINGER-SCIENCE+BUSINESS MEDIA, B.V.

First edition 1992

Originally published by Chapman & Hall in 1992

Typeset in 10/11½ Palatino by Falcon Typographic Art Ltd, Edinburgh

ISBN 978-0-412-37150-9 ISBN 978-94-011-2374-7 (eBook)
DOI 10.1007/978-94-011-2374-7

A catalogue record for this book is available from the British Library

Library of Congress Cataloging-in-Publication data available

For Simon David

Contents

Preface

The decision to write a book about the practical aspects of the baculovirus expression system stems from the numerous phone calls for help we have had, and from the many visitors to our laboratories requiring assistance to find the elusive polyhedrin-negative virus containing their favourite gene. We have also organized two expression system workshops and from the manuals we wrote for these, it seemed a logical progression to extend them into book form. We appreciate that those who are 'old-hands' at the baculovirus expression system may have differing views on some of our procedures, but the methods in this book are presented in the light of our own experiences in the laboratory and from our practical workshops, and we hope that the book will be especially useful to those new to the system.

The first three chapters give the background information to the baculovirus expression system, and includes advice on how to choose the right transfer vector and discusses the various methods that are available to select recombinant viruses. The practical chapters concentrate on those aspects which are novel to the baculovirus system (insect cell culture, virus amplification and titration, etc.) and, in general, leave the standard molecular biological techniques to the other excellent laboratory manuals that are available. However, for completeness sake and to avoid constant reference to other manuals, we have included brief details of some standard techniques where they are integral to the success of the baculovirus protocols.

Many people have wittingly, or unwittingly, contributed to the preparation of this book. We cannot thank them all by name but wish them to know that we are grateful for the help received. In particular we would like to thank the past and present members of our research groups; at the Polytechnic, Susan Mann, Alison Lawrie, Dr Susan Marlow, Kate Trapnell, Dr Matthew Weitzman, Dr Allan Atkinson, Louis Obosi, David Hughes, Chris Palmer, Kirsti Joyce,

Davin Miller and Dr Jason Pickering; and at IVEM, Dr Ulrike Weyer, Dr Paul Kitts, Dr Alison Clarke (née Merryweather), Dr Lorna Stewart, Dr Katy Gearing, Dr Stuart Knight, Steve Howard, Martin Ayres, Jan Pullen, Rachael Hawtin and Dr Miguel Lopez-Ferber. Many of the techniques described in this book have been 'tried and tested' by the above. We would also like to thank Karin Chaloner at IVEM for passing on her expertise in preparing insect cell culture media. We would particularly like to acknowledge the inspiration of Paul Kitts and Martin Ayres, who provided the insight for the linear DNA technique described in Chapter 4. This methid, more than any other, has revolutionized the production of recombinant baculoviruses.

We would also like to thank Derek Whitely of Oxford Polytechnic for his excellent line drawings (Figures 1.1, 1.2, 5.1 and 5.3) and for preparing the drawing on the front cover, and Barbara Southall, Susan Marlow and Chris Hatton for help with some of the photography.

We are grateful to Dr Susan Hemmings and Philippa MacBain, and others in the sub-editorial and production departments at Chapman & Hall for their encouragement and guidance, and for keeping us just about on time.

Finally, we are grateful to our son, Simon, for being a well-behaved baby and allowing us to continue writing and editing!

Linda A. King
Robert D. Possee

The baculoviruses 1

1.1 INTRODUCTION

The object of this chapter is to provide a brief overview of the *Baculoviridae*. This book is a practical guide to the baculovirus expression system and it is inappropriate to attempt a comprehensive review of all aspects of these viruses. Here, we will describe those topics which are important to an understanding of the expression system. Other published articles provide detailed accounts of the use of baculoviruses as insecticides (Entwistle and Evans, 1985; Podgwaite, 1985; Huber, 1986), summaries of their biology (Mazzone, 1985; Granados and Williams, 1986; Kelly, 1987; Volkman and Keddie, 1990) and reviews of their molecular biology (Blissard and Rohrmann, 1990) and their use as expression vectors (Luckow and Summers, 1988a; Miller, 1988; Maeda, 1989a; Atkinson *et al.*, 1990a; Bishop and Possee, 1990). A description of the use of baculoviruses as expression vectors will be given in Chapter 2.

1.2 ISOLATION AND HOST RANGE

Baculoviruses have only been isolated from invertebrates. Most examples have been found in insect species, but there are some reports of baculoviruses which are pathogenic for crustacea (Couch, 1974; Summers, 1977). Baculovirus infections have been described in over 600 species of insects including Lepidoptera (butterflies and moths), Hymenoptera (sawflies and wasps), Diptera (flies), and Coleoptera (beetles) (Martignoni and Iwai, 1986). The devastating effects that they can have on natural populations of insects have made them an obvious choice for use as biological agents to control pest insects (Entwistle and Evans, 1985). For many years the major interest in baculoviruses was for this very purpose; it is only in the last decade that the development of the baculovirus expression vector system has overshadowed, but not replaced, the original reason for studying these viruses. If chemical insecticides had not been so successful in cheaply controlling insect pests, then baculoviruses might have played a more prominent role in agriculture in the last 100 years. At the end of this chapter we will describe how the development of the expression vector system may assist the further exploitation of these viruses as insecticides.

It is important to stress that baculovirus infections have not been reported in humans, mammals or other vertebrate or invertebrate species. Furthermore, they do not infect plants. These are important considerations for those contemplating setting up the expression system in their own laboratories. Baculoviruses may be considered as one of the safest expression systems available at the present time.

1.3 STRUCTURE AND CLASSIFICATION

Baculoviruses have a large, double-stranded, covalently-closed, circular DNA genome of between 88 and 200 kbp (Arif, 1986). This is associated with a highly basic (arginine-rich) protein of 6.5 kDa (Tweeten *et al.*, 1980; Kelly *et al.*, 1983; Wilson *et al.*, 1987), within a rod-shaped nucleocapsid which contains a 39 kDa capsid protein (Thiem and Miller, 1989). Other structural components almost certainly remain to be identified. The size of the virus genome determines the length of the nucleocapsid, which may be 200–400 nm. The width remains constant at about 36 nm (Fraser, 1986).

The nucleocapsids are further packaged within a lipoprotein envelope to form the virus particle or virion. These structures may be occluded within a crystalline matrix or polyhedron (see Figure 1.1), consisting largely of a single protein (polyhedrin) of about 30 kDa (reviewed by Rohrmann, 1986), and form in the nucleus of infected cells. Polyhedra are large structures, ranging in size from 1 to 15 μM in diameter with an outer polysaccharide envelope which appears to confer additional strength and protection (Vlak *et al.*, 1988; Williams *et al.*, 1989).

Variation in the number of nucleocapsids within each virus particle packaged by the polyhedron provides a convenient system for the classification of these viruses. The baculoviruses are divided into three subgroups.

Subgroup A, or nuclear polyhedrosis viruses (NPV), have several virus particles occluded by the polyhedrin protein to form polyhedra. According to the particular isolate, there may be single (S) or multiple (M) nucleocapsids contained by each virus particle. An example of an

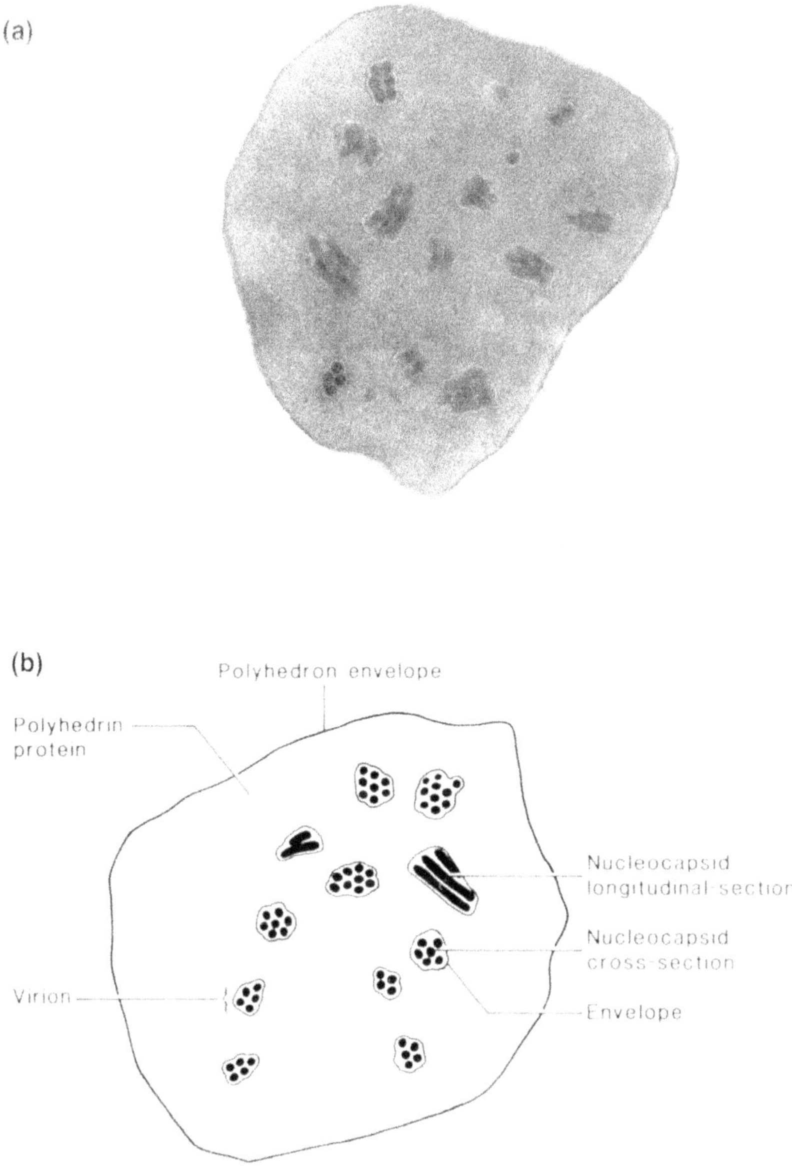

Figure 1.1 Electron micrograph (a) and representative sketch (b) of a section of a polyhedron of AcMNPV. The virus nucleocapsids are clearly visible within the enveloped virus particles embedded within the polyhedrin protein. The polyhedron is surrounded by an electron dense polyhedron envelope.

MNPV is shown in Figure 1.1. It is the viruses from this category that have been developed as expression vectors.

Subgroup B, or granulosis viruses (GV) have single virus particles, containing only one nucleocapsid, occluded by the granulin protein to form granules (reviewed by Moore *et al.*, 1987).

Subgroup C viruses do not produce occlusion bodies and are referred to as non-occluded viruses (NOV) (reviewed by Moore *et al.*, 1987).

Baculoviruses are usually named after the host from which they were isolated. While this system is convenient, it ignores the genetic relatedness of viruses from different species. For example, the baculovirus isolated from the alfalfa looper was designated *Autographa californica* (Ac) MNPV (Vail *et al.*, 1971). However, baculoviruses which are almost identical to AcMNPV have been found in *Trichoplusia ni* (Heimpel and Adams, 1966), *Galleria mellonella* (Smith and Summers, 1979) and *Rachiplusia ou* (Summers and Anderson, 1973).

The AcMNPV has been extensively studied at the molecular level, largely because of its efficient replication in cell culture (see section 1.5). In consequence, it was the logical example to be exploited as an expression vector and forms the major topic of discussion in this book. It should also be mentioned, however, that the *Bombyx mori* (silkworm) (Bm) NPV has also been developed as an efficient expression vector (Maeda *et al.*, 1985; Maeda, 1989a). This virus system, while particularly useful for producing recombinant proteins in silkworm larvae, which are easily reared and handled, has not achieved such widespread popularity as the AcMNPV system.

1.4 BACULOVIRUS REPLICATION *IN VIVO*

The larval or caterpillar stage of the insect life cycle is the most susceptible to infection with NPVs (Figure 1.2). Polyhedra are ingested when the insect feeds on contaminated diet and dissolve in the alkaline environment of the mid-gut to release the virus particles. After negotiating the peritrophic membrane lining the gut, the virus lipoprotein envelope then fuses with the plasma membrane of the gut wall cells and liberates nucleocapsids into the cytoplasm. The nucleocapsids serve to transport the virus DNA to the nucleus of the cell; it is unclear whether the nucleocapsid enters the nucleus or merely 'injects' the virus DNA via a nuclear pore.

The processes of virus gene expression have been most thoroughly examined using cultures of insect cells maintained *in vitro* and will be described below (see section 1.5). What was apparent from the earliest studies on virus replication *in vivo* was the fact that baculoviruses produce two distinct structural forms in a bi-phasic replication-cycle. In the infected gut cells, nucleocapsids are formed by about 8 hours post-infection (h p.i.) and begin to bud through the nuclear membrane by 12 h p.i., thus acquiring a lipid envelope. This membrane appears to be 'lost' in the cytoplasm, but the nucleocapsid gains another as it buds through the plasma membrane. In the course of this latter process, it also acquires a virus-encoded glycoprotein of 67 kDa (gp67; Whitford *et al.*, 1989), which has been inserted into the plasma membrane. This protein most probably serves to attach the budded virus to other susceptible cells within the insect; in cell culture the budded virus is 1000-fold more infectious than

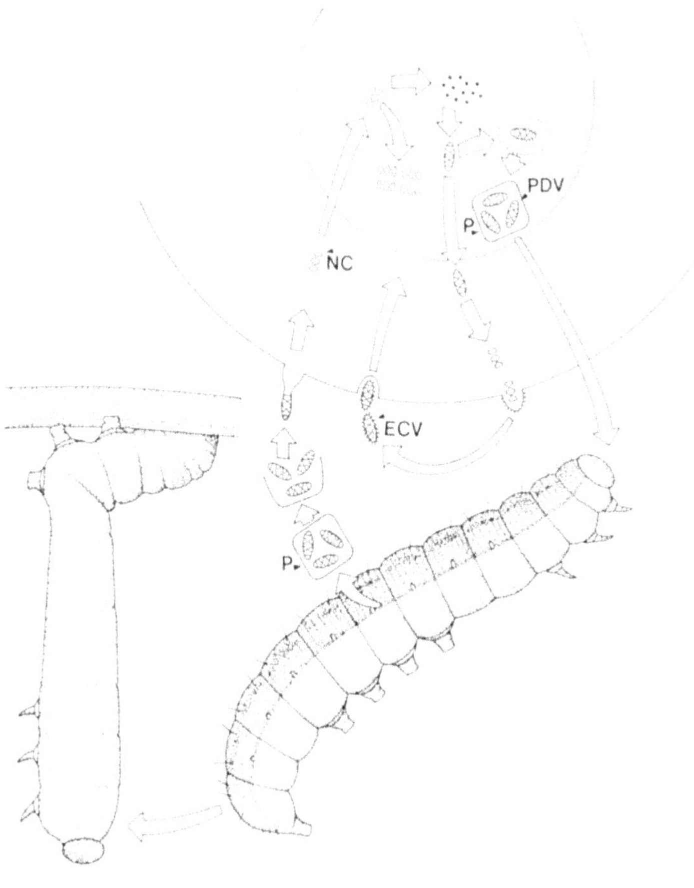

Figure 1.2 Diagrammatic representation of the life cycle of a baculovirus in insect larvae and the insect cell (P, polyhedra; ECV, budded extracellular virus with gp67; PDV, polyhedra-derived virus; NC, nucleocapsid).

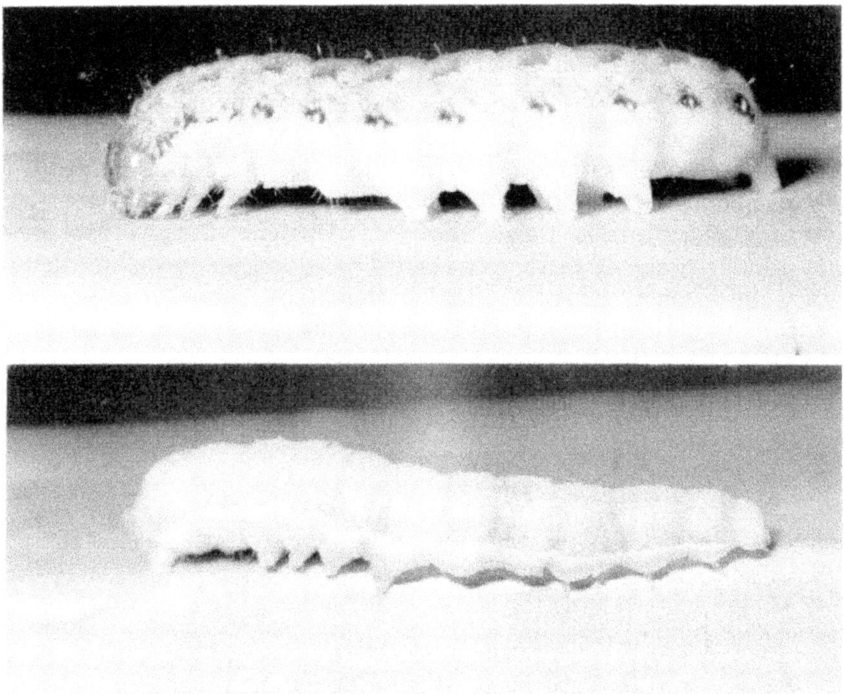

(a)

(b)

Figure 1.3 Photographs comparing (a) the general appearance of non-infected and (b) *Mamestra brassicae* MNPV-infected *M. brassicae* larvae (7 days post-infection). Similar effects are observed in other MNPV-infected larvae.

virus particles released from polyhedra, which lack gp67 (see below) (Volkman *et al.*, 1986).

The budded, or extracellular virus (ECV) is released into the haemolymph to infect other cells and disseminate infection throughout the insect; affected tissues include fat bodies, nerve cells and haemocytes. The cells infected in the second round of virus replication in the insect larva also produce ECV, but in addition occlude virus particles within polyhedra, in the nucleus (Figure 1.2). The virus particles occluded within polyhedra, which are genetically identical with the ECV, obtain their lipid envelope *de novo* within the nucleus and lack the gp67 found in the budded virus phenotype. The accumulation of polyhedra within the insect proceeds until the host consists almost entirely of a bag of virus (Figure 1.3). In the terminal stages of infection the insect liquifies and thus releases polyhedra which can infect other insects. Volkman and Keddie (1990) aptly described the infected insect at this stage as an 'amorphous puddle'.

In summary, baculoviruses produce two structurally distinct forms of virus. The first, the ECV, serves to spread the infection within an individual and the second, the polyhedron, acts as the dispersal agent between individuals and as a survival mechanism when susceptible hosts are not available.

1.5 BACULOVIRUS REPLICATION *IN VITRO*

It is relatively easy to establish continuous cultures of insect cells *in vitro*. The starting material may include pupal ovarian tissue or fat bodies, haemocytes and other organs from larvae or homogenates of entire larvae. The advantage of using specific tissues is that one may then derive a range of cell lines which do not simply reflect the fastest growing cell type and may be more useful for propagating virus. Detailed protocols for the maintenance of continuous insect cell lines are provided in Chapter 5.

Cell lines which support the replication of AcMNPV have been derived from *Spodoptera frugiperda* (Sf) (Fall army worm) pupal ovarian tissue (Vaughn *et al.*, 1977) or ovaries from adult *T. ni* (cabbage looper) (Hink, 1970) and *Mamestra brassicae* (cabbage moth) (Miltenberger, 1979; King *et al.*, 1991). Although other baculoviruses have been grown in cell culture, few match the efficiency of the AcMNPV–Sf cell combination.

The study of baculovirus replication *in vitro* has greatly simplified experiments to understand the kinetics of virus gene expression and replication. Furthermore, it was an essential prerequisite for the development of the baculovirus expression vector system. Virus infection may be established *in vitro* most conveniently by harvesting haemolymph, containing infectious virus particles (ECV), from infected larvae and using it to inoculate susceptible cells in culture. The virus particles which subsequently bud from the plasma membrane of these cells, into the culture medium, may be used to amplify virus stocks further. These procedures are described in Chapters 6 and 10.

It is rarely necessary, however, to perform the primary operation of inoculating cultured cells with haemolymph from virus-infected larvae; stocks of AcMNPV amplified *in vitro* are widely available from many laboratories. Some purists, however, are wary of using viruses which have been passaged repeatedly in cell culture. It is well documented that AcMNPV can suffer insertions of host cell DNA and transposable elements within its genome (reviewed by Blissard and Rohrmann, 1990). A consequence of one such insertion within a gene encoding a 25 kDa protein, is the production of viruses which yield few polyhedra (FP phenotype) (Beames and Summers, 1988, 1989). Clearly, baculovirus genomes, in common with those of other viruses, are subject to mutations which may produce altered phenotypes. Fortunately, it is our experience that these potential problems have had a negligible effect on the success of the baculovirus expression system. It is quite rare for recombinant baculoviruses to undergo mutations which affect recombinant protein expression. The potential for the accumulation of genotype changes may be minimized by exercising good practice in propagating these viruses (see Chapters 6 and 7).

1.5.1 BACULOVIRUS GENE EXPRESSION AND REPLICATION

The bi-phasic production of ECV and polyhedra observed in insect larvae is also found in cell culture and, in general, the processes are similar. In this section more details are provided of the molecular events accompanying virus replication. The budded or ECV form of AcMNPV enters insect cells in culture by the process of adsorptive endocytosis. The nucleocapsids serve to translocate the DNA to the nucleus of the cell where virus replication is initiated. Baculovirus genes are expressed in a regulated fashion in infected insect cells. For convenience, and to match the effects of inhibitors of virus replication, virus gene expression is divided into four phases. These are: immediate-early (α); delayed-early (β); late (γ); very late (δ). In general, the expression levels attained in each succeeding phase is higher than that of the preceding one (represented diagrammatically in Figure 1.4). A brief review of the four phases is included below. A representation of the virus replication cycle within infected cells is shown in Figure 1.2 and in the scanning (Figure 1.5) and transmission (Figure 1.6) electron micrographs of an infected cell.

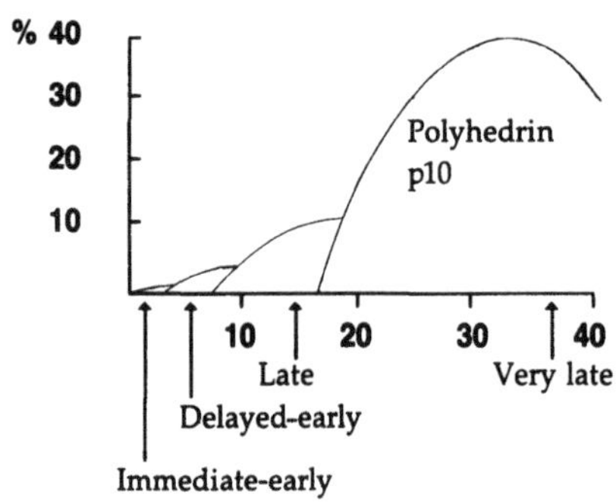

Figure 1.4 Schematic representation of the four phases of baculovirus gene expression *in vitro*. The levels of gene expression are not drawn to scale.

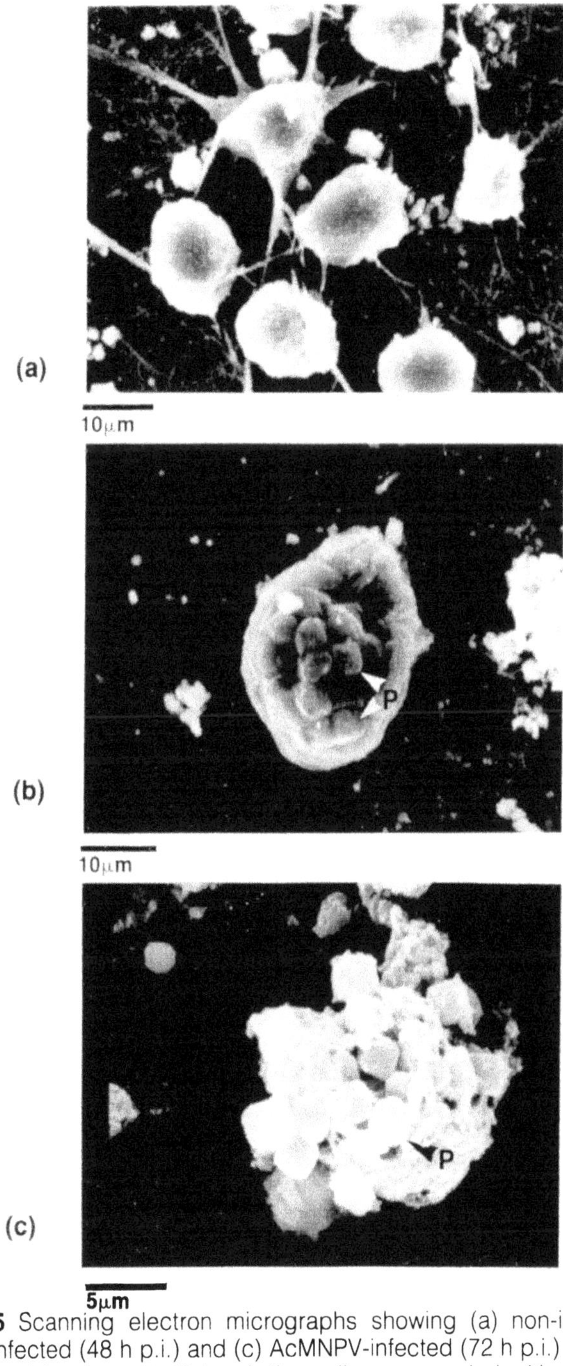

Figure 1.5 Scanning electron micrographs showing (a) non-infected, (b) AcMNPV-infected (48 h p.i.) and (c) AcMNPV-infected (72 h p.i.) *M. brassicae* cells cultured *in vitro*. At 48 h p.i. the cells are rounded with an enlarged nucleus packed with polyhedra (P). At 72 h p.i. the cell lyses to release the polyhedra. Scale-bars are shown below each panel. We thank Dr S.A. Marlow for supplying this figure.

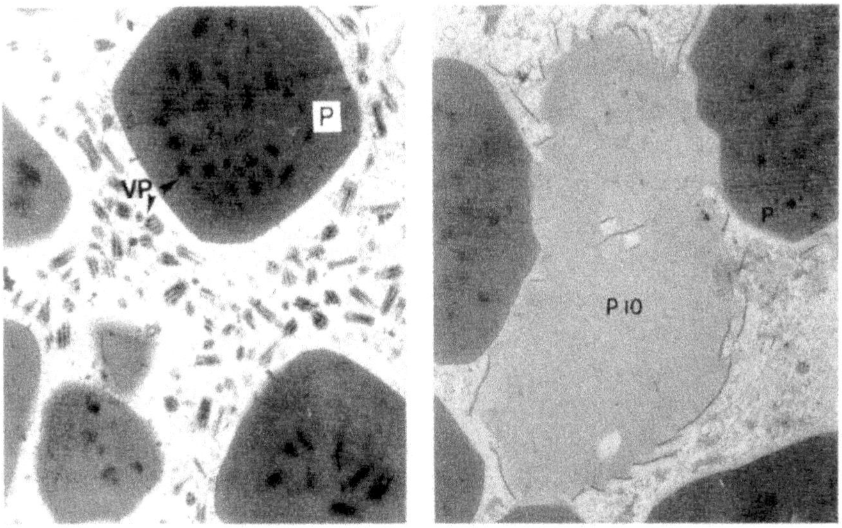

Figure 1.6 Transmission electron micrographs of sections through AcMNPV-infected *M. brassicae* cells at 48 h p.i., showing polyhedra (P), polyhedra-derived virus particles (VP), before and after occlusion, and p10 protein (P10).

(a) Immediate-early genes (α)

Immediate-early genes are defined as those genes which can be transcribed in the presence of inhibitors of protein synthesis (e.g. cycloheximide) (Kelly and Lescott, 1981; Guarino and Summers, 1986a,b). Two examples have been described to date. These are IE-1 (Guarino and Summers, 1986a) and IE-N (Carson *et al.*, 1988). Their classification as immediate-early genes has been confirmed by the fact that copies of these genes, inserted into plasmids, are transcriptionally active after transfection into uninfected insect cells. Evidently, these virus genes are transactivated by host transcriptional factors and do not require *de novo* synthesis of other virus gene products.

(b) Delayed-early genes (β)

An early study using cycloheximide and other inhibitors of protein synthesis was instrumental in defining the second phase of gene expression (Kelly and Lescott, 1981). When cells were treated with these inhibitors and then returned to normal growth conditions, a distinctive pattern of protein expression was observed. Some proteins were expressed immediately after the inhibitors was removed, while others were produced after a delay; hence the division between immediate-early and delayed-early gene products.

These results have been extended in more recent experiments where immediate-early gene products have been found to trans-activate the expression of delayed-early genes after transfection of uninfected insect cells (Guarino and Summers, 1986a; Carson *et al.*, 1988). The 39K delayed-early gene is first detected in infected cells between 3 and 6 h p.i. The chloramphenicol acetyl transferase (CAT) gene coding sequences were inserted, in frame, with a truncated AcMNPV 39K gene in a recombinant plasmid. Transfection of insect cells with this construct did not result in detectable CAT activity. When a plasmid containing the IE-1 gene was co-transfected with the 39K-CAT construct, however, significant quantities of CAT enzyme activity were produced (Guarino and Summers, 1986a). It was concluded from these experiments that the immediate-early class of gene products is required in the virus infection to transactivate the delayed-early genes. It has also been demonstrated that IE-N can augment the transactivation of 39K in combination with IE-1, but not when transfected with 39K alone (Carson *et al.*, 1988).

These gene transactivation experiments have all been conducted with the virus genes inserted into plasmids and subsequently transfected into insect cells; it remains to be seen whether the conclusions drawn from these results will apply to the control of gene expression in the intact virus genome. However, these results, and those for the IE genes, have significance for the future developments of baculovirus expression vectors (see Chapter 2).

(c) Late genes (γ)

The third class of virus genes expressed in infected cells coincides with the onset of virus DNA replication at about 6 h p.i. If virus DNA synthesis is inhibited with aphidicolin, the late genes are not transcribed (Miller *et al.*, 1981; Wang and Kelly, 1983). The virus encodes a DNA polymerase gene, the transcription of which is also inhibited by aphidicolin (Tomalski *et al.*, 1988). Virus genes which are expressed during this phase include those encoding structural elements of the virus particles, e.g. the basic protein (Wilson *et al.*, 1987), the capsid protein (Thiem and Miller, 1989) and the virus membrane glycoprotein (gp67) (Whitford *et al.*, 1989). Transcription of these genes is probably mediated by the action of an unusual RNA polymerase which is α-amanitin resistant and induced in cells late in infection (Grula *et al.*, 1981; Fuchs *et al.*, 1983). It is unclear whether this enzyme is virus-encoded or simply a modification of a host cell RNA polymerase. Interestingly, the delayed-early 39K gene promoter is also active in the late phase of baculovirus gene expression. By virtue of a shift in the position of transcription initiation to a site about 25 nucleotides upstream from the early mRNA start site, the gene is expressed at least until 18 h p.i. (Guarino and Summers, 1986a).

(d) Very late genes (δ)

There is some debate as to whether this class of genes should be regarded as a division separate from the preceding late genes. The principal reason for designating a fourth class is that they are transcribed in the period when the virus is assembling occlusion bodies within the nucleus of the infected cell, from about 15 h p.i. The very late gene products include the polyhedrin protein, which forms the matrix of the occlusion body, and the p10 protein, which most probably has a role in polyhedra formation (Vlak *et al.*, 1988; Williams *et al.*, 1989). The p10 protein forms a crystalline matrix in the infected cell nucleus (see Figure 1.6) that is associated with polyhedra formation, although it does not form part of the mature polyhedron.

Another justification might be to define very late genes as those which play no role in the formation of infectious virus particles; the polyhedrin and p10 genes may be deleted from the virus genome without affecting virion (ECV) production (Smith *et al.*, 1983; Vlak *et al.*, 1988). These two very late genes have certainly been the major focus for development of baculovirus expression vectors, since their promoters are extremely efficient and can result in their combined proteins accounting for up to 50% of the total cell protein mass in the terminal stages of infection.

1.5.2 BACULOVIRUS GENE PROMOTERS

Our gradually increasing knowledge of baculovirus gene promoters has paralleled the development of the expression vector system. Many studies which have reported new expression vectors have added information concerning the nature of the promoters. In this section we will simply be concerned with the data pertaining to the primary sequence requirements of the baculovirus promoters.

The interest in the use of the very late gene promoters (polyhedrin and p10) as expression systems has focused attention on these structures. The sequence of these two promoters in AcMNPV is shown in Figure 1.7. Transcription is initiated at a TAAG motif located 50 (polyhedrin; Howard *et al.*, 1986) or 70 (p10; Kuzio *et al.*, 1984) nucleotides upstream from the respective translation initiation codons. This motif is found in all of the baculovirus late and very late gene promoters identified to date (Rohrmann, 1986; Blissard and Rohrmann, 1990).

The importance of the 5′ non-coding leader sequence of the polyhedrin gene was suggested by early results from Smith *et al.* (1985), which showed that removing part of this sequence decreased the level of expression. These results were extended in a more detailed study (Matsuura *et al.*, 1987) where progressive deletions were made between the translation initiation codon and the transcription start site (i.e. 3′ to 5′). These data showed that the length of the 5′ non-coding leader could be correlated directly with the level of expression. The seven nucleotides before the ATG codon were found to be particularly important for maximum promoter activity. A series of linker-scan mutations in the 5′ non-coding leader also confirmed the importance of this region (Rankin *et al.*, 1988; Ooi *et al.*, 1989). Furthermore, these latter studies also demonstrated that replacing eight nucleotides spanning the TAAG motif with a synthetic oligonucleotide linker resulted in a 2000-fold decrease in promoter activity. The transcription start site was further dissected in a study by Gearing and Possee (1991), where point mutations were introduced into the 11 nucleotides spanning the TAAG motif. Alteration of the TAAG sequence abolished promoter activity, while changes in the flanking regions only resulted in a small decrease in

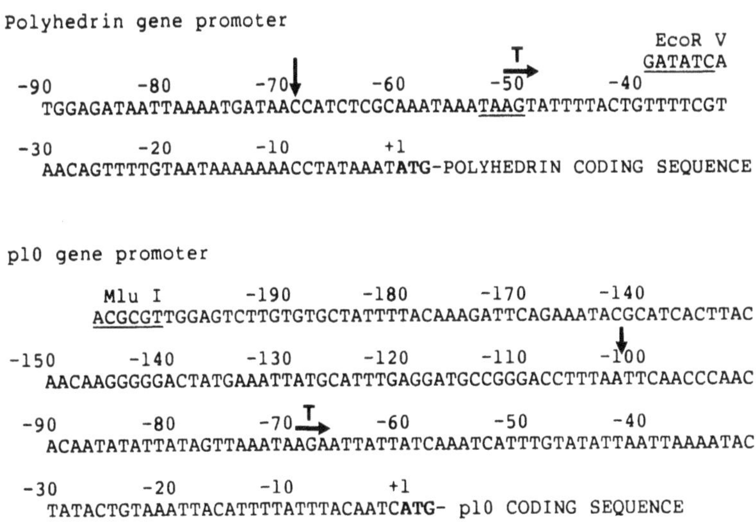

Figure 1.7 Nucleotide sequences of the AcMNPV polyhedrin and p10 gene promoters, presented to show transcription initiation sites (horizontal arrow and T) and the extent of the promoter regions (vertical arrows).

expression. The TAAG motif appears to be a universal feature of all the late and very late baculovirus gene promoters; it also serves as the late transcription start site in the delayed early 39K gene promoter (for a review, see Blissard and Rohrmann, 1990).

The region upstream from the transcription start site has also been analysed using deletion mutants (5′ to 3′) (Possee and Howard, 1987). It was concluded that between 7 and 20 nucleotides were required for maximum promoter activity. Interestingly, Ooi *et al.* (1989) reported that inserting a synthetic linker between 12 and 22 nucleotides upstream from the TAAG sequence increased the levels of steady-state mRNA by up to 50%. Both studies confirm that the sequences upstream from the mRNA start site are relatively unimportant.

Similar results have been obtained with the AcMNPV p10 gene promoter (Weyer and Possee, 1988, 1989). The conclusion from these studies is generally the same as with the studies of the polyhedrin promoter. The p10 promoter consists of about 100 nucleotides extending upstream from the translation initiation codon. There is an absolute requirement for the 5′ non-translated leader sequence and about 30 nucleotides upstream from the transcription start site. The p10 promoter is fully functional when located in a heterologous location within the virus genome, namely, upstream of the polyhedrin gene (Weyer *et al.*, 1990).

Other virus gene promoters have been less well characterized at the primary sequence level. The IE-1 gene is transcribed from two mRNA start sites, implying that it has two promoters (Chisholm and Henner, 1988). The first of these structures appears to be functional between 0 and 2 h p.i. and produces a spliced transcript of 2.1 kb. The 5′ end of this transcript maps to a position about 4 kbp upstream of the mRNA start site of the second, 1.9 kb transcript. The 1.9 kb transcript reaches its steady-state level 30 min after the virus adsorption period and maintains this level throughout the infection (Chisholm and Henner, 1988). The 39K gene promoter also utilizes two mRNA start sites in the delayed-early and late phases of gene expression (see above). The DNA polymerase gene also has two start sites for mRNA initiation, but these are both active in the early phase of gene expression (Tomalski *et al.*, 1988).

Another interesting feature of immediate-early/delayed-early gene expression is the role of enhancer elements. In AcMNPV these consist of five regions of homologous repeats (hr1–hr5) containing repeated sequences with multiple *Eco*RI sites (Cochran and Faulkner, 1983; Guarino *et al.*, 1986b). When linked in *cis* with immediate-early and delayed-early genes, these sequences may enhance transcription by up to 1000-fold (Guarino and Summers, 1986b; Nissen and Friesen,

1989). The role, if any, of these enhancer elements in late and very late gene transcription remains to be elucidated.

1.6 GENETIC ENGINEERING OF BACULOVIRUS INSECTICIDES

In a book devoted to the use of baculoviruses as expression vectors, a section on their use as insecticides may seem misplaced. However, the expression system may be of considerable use in modifying the effectiveness of the natural viruses Unmodified baculoviruses have been used successfully as insecticides, but suffer the disadvantage of a relatively slow mode of action in comparison with chemicals. This has discouraged their widespread use in agriculture. It is proposed that the expression of insecticidal proteins in the target host, via the use of baculovirus expression systems, may reduce the time taken to kill the larva. For example, expression of the *Heliothis virescens* juvenile hormone esterase in *T. ni* larvae infected with a recombinant AcMNPV, reduced the weight gain of the insects (Hammock *et al.*, 1990). The *Bacillus thuringiensis* delta endotoxin, produced by a recombinant AcMNPV, reduced the feeding of those insects which ate a contaminated diet (Merryweather *et al.*, 1990). Expression of a diuretic hormone by the silkworm baculovirus (BmNPV) reduced the time required by the virus to kill the host, although only by 20% (Maeda, 1989b).

Clearly, although the primary use of baculovirus expression vectors has been to produce recombinant proteins *in vitro* for research, medical and veterinary purposes, the new technology may assist the development of improved baculovirus insecticides. For those scientists with an interest in this area we have included a description of the methods for propagating baculoviruses *in vivo* and assessing the infectivity of these viruses in insects (Chapter 10). We would encourage the testing of recombinant baculoviruses in insect larvae because it is likely that some of the proteins expressed for quite different reasons may fortuitously have insecticidal properties.

Chapter 10 will also serve to introduce the concept of using insect larvae as a system for the expression of foreign gene products *in lieu* of cell culture techniques. This method has been suggested before, and tested with AcMNPV and *T. ni* insects (Overton *et al.*, 1987), but has yet to be significantly exploited.

The development of baculovirus expression vectors 2

2.1 INTRODUCTION AND HISTORICAL PERSPECTIVE

The major attraction of baculoviruses as an expression vector system was originally the virus-encoded polyhedrin gene. This produces large amounts of polyhedrin protein (28–30 kDa) in virus-infected insect cells in the latter stages of the replication cycle (see Chapter 1). The protein is required in the course of a normal infection cycle to package virus particles within occlusion bodies or polyhedra, which protect the virus particles in the environment between susceptible hosts. Although polyhedra are required to infect insects *per os*, their production is not necessary to maintain an infection in cultured cells *in vitro*. The redundancy of the polyhedrin protein was demonstrated by making deletions in the coding sequence which did not affect the synthesis of infectious, non-occluded virus particles (ECV) in the late phase of the infection cycle, in cell culture (Smith *et al.*, 1983a). It was then a logical step to replace the polyhedrin coding region with foreign gene sequences and thus derive recombinant protein from the polyhedrin promoter. Polyhedrin-negative viruses can be propagated *in vivo* by injecting ECV directly into the haemolymph (see Chapter 10).

Similar experiments also demonstrated that the p10 gene coding sequences could be deleted without affecting ECV production. Viruses that lack the p10 gene, but which retain the polyhedrin gene, are still able to make polyhedra, although they appear to be more fragile than when the p10 protein is synthesized (Vlak *et al.*, 1988; Williams *et al.*, 1989).

The first reports of the use of baculoviruses as expression vectors were published by Smith *et al.* (1983) and Pennock *et al.* (1984) who used AcMNPV to produce β-interferon and β-galactosidase, respectively in Sf cells. Maeda *et al.* (1985) also expressed β- interferon using the BmNPV system in the silkworm. In the succeeding years a wide variety of recombinant proteins have been expressed using the system (see Chapter 3).

In the early days of its development, the baculovirus system suffered the disadvantage of a lack of detailed knowledge concerning the nature of the very late promoters utilized in the expression vectors. This resulted in the use of polyhedrin promoter constructs in some vectors which were sub-optimal for foreign gene expression and produced disappointing results. A further complication has been that many groups have constructed their own, uniquely named versions of essentially the same expression vector. In this chapter we will review some of the expression vectors and provide guidelines for the most appropriate ones to be used in expression studies. The general principles of the system will be explained, including a review of the different methods now available for selecting recombinant viruses. A survey will also be made of the more recent developments in the baculovirus expression system, which have seen the utilization of other virus gene promoters, active in other phases of gene expression, and the development of vectors for the expression of two or more foreign genes.

2.2. THE MERITS OF THE BACULOVIRUS EXPRESSION SYSTEM

The ideal expression system for all recombinant proteins has yet to be developed. Prokaryotic, yeast and mammalian expression vectors are each appropriate according to the nature of the recombinant protein and its final use. The baculovirus expression system, on balance, has considerably more advantages than disadvantages and may be the method of choice in many circumstances.

2.2.1 ADVANTAGES

(a) Dispensible virus gene products: the virus has two very late gene products (polyhedrin and p10) which are not required for the continued production of infectious virus particles in cell culture. The virus coding sequences may, therefore, be replaced with foreign genes.

(b) Strong gene promoters: the polyhedrin and p10 genes have strong promoters which result in the synthesis of large quantities of proteins in infected cells. Other virus promoters, although less active, are also useful for foreign gene expression.

(c) Temporal factors: the expression of the very late virus gene promoters occurs in cells after the maturation of budded, infectious virus particles. Consequently, if a cytotoxic protein is to be synthesized it will not adversely affect virus replication.

(d) Size of insert: baculovirus genomes are remarkably variable in size (88–200 kbp). The virus can accommodate large amounts of foreign DNA without affecting normal replication and DNA packaging; the nucleocapsid simply extends lengthwise to accommodate the extra DNA.

(e) Post-translational processing: baculoviruses replicate in eukaryotic cells which ensures faithful processing of foreign gene products. Insect cells do, however, show some differences in the pattern of glycosylation in virus-infected cells (see below).

(f) Safety: baculoviruses replicate only in invertebrates; there is no risk to the user of the system.

(g) Scale-up: insect cells are amenable to large-scale volume production in fermenter systems.

2.2.2 DISADVANTAGES

(a) Discontinuous expression: baculovirus infection of insect cells results in the death of the host and hence the necessity to reinfect fresh cultures of Sf cells for each round of protein synthesis. This may be inefficient for production on a commercial scale, although it is acceptable for research purposes.

(b) Glycosylation: the process of glycosylation in insect cells is demonstrably different from the same pathway in vertebrate cells. This will be described in more detail in Chapter 3. In brief, insect cells produce glycoproteins with relatively simple, unbranched sugar side-chains with a high mannose content. The most obvious consequence of this processing is that the glycoproteins produced in insect cells have a greater mobility in denaturing polyacrylamide gels; their molecular weights are measurably different.

2.3 GENERAL PRINCIPLES FOR INSERTING FOREIGN GENES INTO THE BACULOVIRUS GENOME

A consequence of the size of the baculovirus genome (AcMNPV, 128 kbp) is that it cannot yet be directly manipulated, using restriction enzymes (RE) and DNA ligases to insert foreign DNA, in a manner analogous to bacterial or yeast vectors. Until recently (see section 2.5.5), unique sites for RE recognition and cleavage were unknown in the AcMNPV genome. Instead, it was necessary to adopt an indirect approach to modify the baculovirus DNA.

Portions of the virus genome spanning the polyhedrin gene were inserted into bacterial plasmids and propagated in *Escherichia coli* hosts. The polyhedrin gene was then manipulated using RE and exonucleases to remove part, or all of the coding region, while leaving the promoter largely intact (Smith *et al.*, 1983a,b; Pennock *et al.*, 1984; Matsuura *et al.*, 1987). (It is sobering today to consider that the considerable effort required to produce the deletion mutants of the polyhedrin gene could now be replaced by a simple polymerase chain reaction (PCR) experiment!) Provision of a unique RE site just downstream of the promoter facilitated insertion of foreign DNA. This final plasmid construction was variously designated as the transfer vector (Smith *et al.*, 1983b) or the transplacement vector (Pennock *et al.*, 1984).

The foreign sequences inserted under the control of the polyhedrin promoter are transferred to the virus genome by co-transfection of insect cells with infectious virus DNA and the transfer vector (Figure 2.1). The various methods available for this step will be reviewed and described in Chapter 7. In the course of the resulting infection, the sequences flanking the foreign gene recombine with the homologous sequences in the virus genome and thus replace the polyhedrin coding sequences. This produces a virus genome, with a polyhedrin-negative genotype, which is the expression vector (see Figure 2.1). The recombinant virus is isolated from the progeny of the co-transfection by performing a plaque-assay and using one of a variety of selection methods to distinguish the recombinant- and parental-virus plaque phenotypes (see section 2.5). A summary of the processes involved in constructing a recombinant virus is presented in Figure 2.1. A very similar approach was adopted for the use of the p10 promoter, although with some modifications because of the lack of an easily selectable phenotypic marker (see below).

2.4 BACULOVIRUS TRANSFER VECTORS

One of the most difficult decisions to be made by the newcomer to the baculovirus system can be which transfer vector to use. There are vectors derived from both the polyhedrin and p10 very late gene promoters; these produce the highest concentrations of virus encoded proteins in infected cells. Recently a vector was described which makes use of the basic protein promoter, a late gene regulatory element which is active between 8 and 20 h p.i. (Hill-Perkins and Possee, 1990). Transfer vectors which utilize other virus promoters have also been developed. Vectors have also been described which permit the insertion of two foreign genes simultaneously into the virus genome. A restriction map of the AcMNPV genome is shown

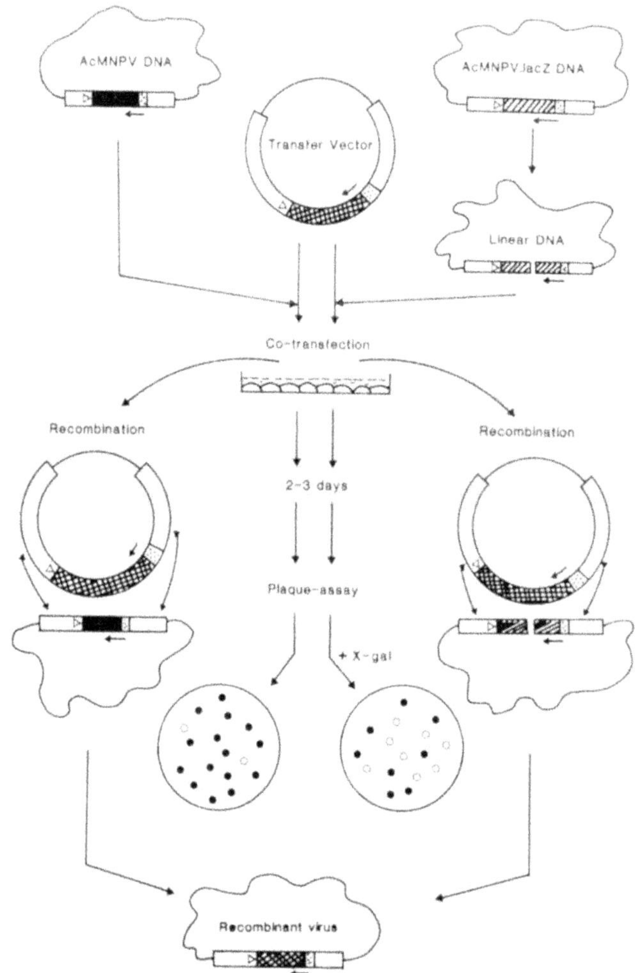

Figure 2.1 Schematic representation of the production of a recombinant baculovirus. The plasmid transfer vector contains the polyhedrin promoter (stippled box) and transcription termination signals (open arrow head) flanked by AcMNPV sequences (open boxes). A foreign gene (cross-hatched box) is inserted under the control of the polyhedrin promoter and then mixed with infectious AcMNPV DNA (or linearized AcMNPV DNA, e.g. AcMNPV.*lac*Z, see text section 2.5.5). Insect cells are co-transfected with the transfer vector and virus DNA using either calcium phosphate co-precipitation, electroporation or lipofection (see Chapter 7). Homologous recombination replaces the polyhedrin gene (solid box) or *lac*Z gene (hatched box) with the foreign gene coding sequences. The progeny virus are screened for polyhedrin- or *lac*Z-negative viruses by plaque-assay or by dot-blot hybridization followed by plaque-assay. Recombinant viruses are identified by the production of polyhedrin- or *lac*Z-negative plaques (open circles) compared with the production of polyhedrin- or *lac*Z-positive plaques (solid circles) by the parental virus.

in Figure 2.2 to depict the positions of the virus genes used as the basis for expression vectors. We will now consider each of the types of transfer vectors in turn.

2.4.1 POLYHEDRIN PROMOTER-BASED EXPRESSION VECTORS

Many expression vectors have been derived from the polyhedrin promoter. Accordingly, this category provides the maximum confusion to the first-time user of the expression system. The situation is complicated further by modifications which are made to the standard transfer vectors by subsequent users, who give alternative names to these 'new' vectors. In consequence it can be almost impossible to discern which transfer vector should be used to express your particular protein of interest. We have provided simple RE maps of the most commonly used transfer vectors which are currently available (Figure 2.3a). Any of these vectors may be selected in the confidence that they will produce the maximum amount of recombinant protein. The differences between them are subtle, but

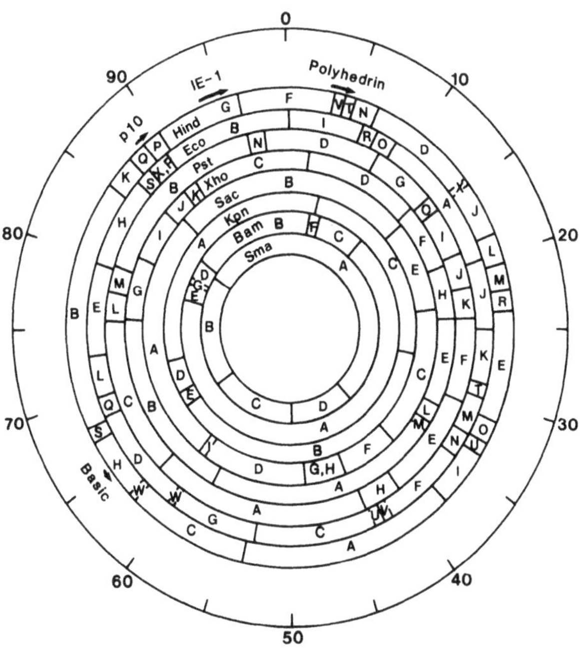

Figure 2.2 Circular restriction enzyme map of AcMNPV to show the loci of the virus gene promoters used in the various expression vectors. The p10, polyhedrin, IE-1 and basic protein genes are shown with the direction of transcription indicated. The outermost circle represents percent of the virus genome. Modified after Vlak and Smith (1982).

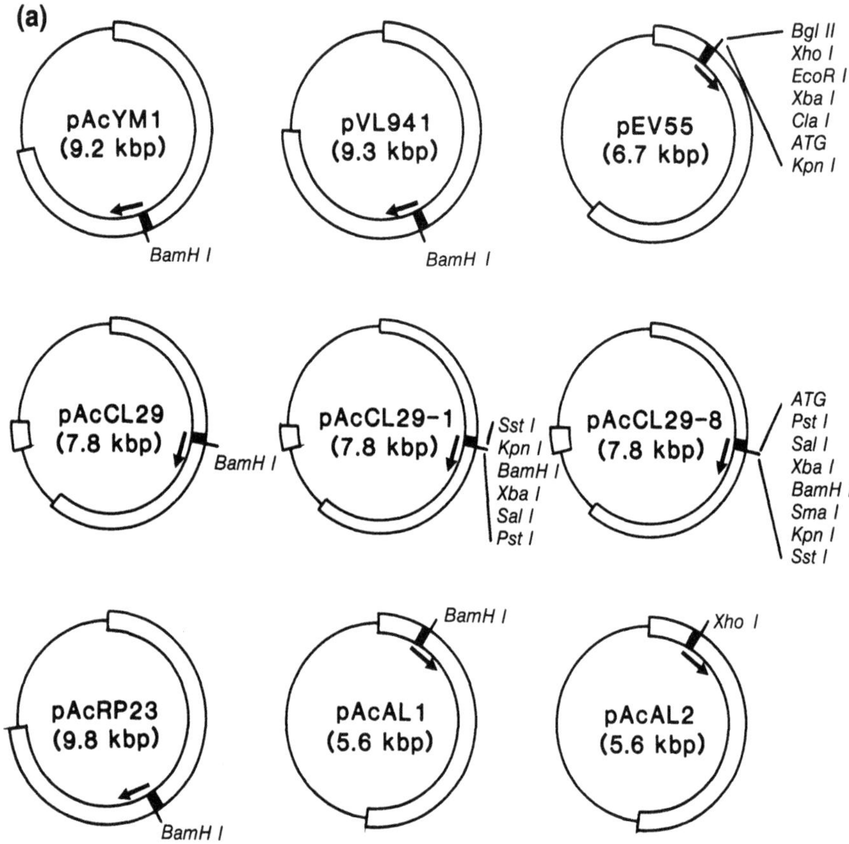

Figure 2.3 (a) Schematic representation of the various types of polyhedrin-based plasmid transfer vectors showing: plasmid size (kbp); location and direction of the polyhedrin promoter (solid box); and the insertion site for the foreign gene coding sequences. pAcYM1 (Matsuura *et al.*, 1987); pVL941 (Luckow and Summers, 1989); pEV55 (Miller *et al.*, 1986); pAcCl29 (Livingstone and Jones, 1989); pAcRP23 (Possee and Howard, 1987); pAcAL1, pAcAL2 (A. M. Lawrie and L. A. King, unpublished data). The pAcCl29 series of transfer vectors contains the M13 intergenic region (open box) and can produce single-stranded DNA in the presence of a helper phage such as M13 K07. The pAcAL1 transfer vector was derived from pAcRP23 by excising a 2874 bp DNA fragment with *Sal*I and inserting it into a modified pT7T3 vector (Pharmacia) containing a *Xho*I site *in lieu* of the polylinker. pAcAL2 was produced by digesting pAcAL1 with *Bam*HI, repairing the ends with the Klenow fragment of *E. coli* DNA polymerase and adding a synthetic *Xho*I linker. A further, more detailed restriction map of pAcYM1 will be found in Chapter 4 (Figure 4.1). (b) Sequences of the polyhedrin promoters in the various plasmid transfer vectors decribed in (a). The gap in these sequences delineates the junction between the polyhedrin promoter and the insertion sites for the foreign gene.

```
                              +1
pAcYM1     AAAAAAACCTATAAATA  CGGATCCGGTTATTAGTACATTTATTAAGCGCTAGATTCTGTG
/pAcCL29                       BamH I


                              -1
pVL941     AAAAAAACCTATAAATATTCCGGATTATTCATACCGTCCCACCATCGGGCG CGGATCC
                                                                BamH I

           TTTCCTGGGACCCGGCAAGAACCAAAAACTCACTCTCTTCAAGGAAATCCGTAATGTTA


                             -1  Bgl II    EcoR I    Cla I
pEV55      AAAAAAACCTATAAAT  AGATCTCGAGAATTCTAGATCGATGGTACCGACTCTGCTGAAGA
                                 Xho I     Xba I      Kpn I

           GGAGGAAATTCTCCTTGAAGTTTCCCTGGTGTCGCTAGATTCTGTGCGTTGTTGATTTACAG


                              +1                  Kpn I          Xba I
pAcCL29-1  AAAAAAACCTATAAATA CGGATCAATTCGAGCTCGGTACCCGGGGATCCTCTAGAGTCGAC
                                        Sst I     Sma IBamH I        Sal I

           CTGCAGGCATGCAAGCTGATCCGGTTATTAGTACATTTATTAAGCGCTAGATTCTGTGCGTT
           Pst I


                              +1                         Sal I       BamH I
pAcCL29-8  AAAAAAACCTATAAATA CGGATCAGCTTGCATGCCTGCAGGTCGACTCTAGAGGATCCCCG
                                               Pst I        Xba I       Sma I
           Kpn I
           GGTACCGAGCTCGAATTGATCCGGTTATTAGTACATTTATTAAGCGCTAGATTCTGTGCGTT
                 Sst I


pAcRP23                      -1
/pAcAL1    AAAAAAACCTATAAAT CCGGATCCTTTCCTGGGACCCGGCAAGAACCAAAAACTCACTC
                              BamH I


                             -1
pAcAL2     AAAAAAACCTATAAAT CCGGATCCCTCGAGGGATCCTTTCCTGGGACCCGGCAAGAACC
                                     Xho I
```

worth considering. The sequences of part of the promoter regions, and the insertion sites of these vectors are shown in Figure 2.3b. In addition, in Chapter 4 we provide a full RE map of pAcYM1 (Figure 4.1), one of the most commonly used vectors.

(a) pAcYM1, pEV55, pAcRP23 and pVL941

The pAcYM1 (Matsuura *et al.*, 1987), pEV55 (Miller *et al.*, 1986) and pAcRP23 (Possee and Howard, 1987) transfer vectors all have the complete 5′ non-coding leader sequence of the polyhedrin gene, which is essential for high-level expression from this promoter (see Chapter 1). pAcYM1 also has the first nucleotide of the translation initiation codon. pVL941 (Luckow and Summers, 1989) has the complete 5′ non-coding region, but also retains the first 34 nucleotides of the polyhedrin gene coding sequence; the ATG has been altered to ATT to prevent translation initiation before the foreign coding sequence is encountered by the host cell ribosome. We do not know of any evidence that these slight variations in the 5′ non-coding leader sequence affect foreign gene expression in baculovirus-infected cells.

The pAcYM1 and pVL941 transfer vectors both have single *Bam*HI insertion sites for foreign gene coding sequences. pEV55 has a polylinker region that also incorporates an ATG codon, which is useful for inserting foreign coding sequences that do not have their own translation initiation codon. Other vectors, not listed here, also retain ATG codons, with varying proportions of the polyhedrin coding region for the production of fusion proteins containing the amino terminal portion of the polyhedrin (e.g. pAc360; Luckow and Summers, 1989). These vectors are, however, becoming redundant with the advent of PCR, which facilitates easy modification of coding sequences to incorporate a translation initiation codon. They may still find uses, however, when expressing small peptides which, in the native form are unstable in insect cells, e.g. a scorpion neurotoxin (Carbonell *et al.*, 1988).

(b) pAcCL29-series and pAcAL1–2

The pAcCL29-series of transfer vectors (Livingstone and Jones, 1989) are very similar to those described above. They were originally derived from pAcYM1. The major difference is that each vector has the M13 intergenic region necessary for single-stranded DNA production, after superinfection of transformed bacteria with an M13 helper phage (e.g. M13 KO7). This is particularly advantageous if modifications need to be made to the inserted foreign gene coding sequences after the initial construction has been made. Single-stranded DNA may be produced and used in site-directed mutagenesis techniques with synthetic oligonucleotides. The pAcCL29 transfer vectors are also slightly smaller than pAcYM1 and should, therefore, be able to accept more foreign DNA before becoming unstable.

The vectors pAcAL1–2 were derived from pAcRP23 and are similar to the pAcCL29-series, in that they also have single-stranded capability (based on pT7T3), but these vectors are smaller and have been shown to be useful in accommodating large foreign gene coding-sequences (A. M. Lawrie and L. A. King, unpublished data).

(c) pJV*Nhe*I and pAcDZI

Another interesting and useful group of polyhedrin promoter-based transfer vectors are those which also have a copy of the *E. coli lac*Z coding sequences, under the control of another promoter, inserted upstream of the polyhedrin gene. These include pJV*Nhe*I (Vialard *et al.*, 1990) and pAcDZ1 (Zuidema *et al.*, 1990). In pJV*Nhe*I the *lac*Z sequences are under the control of a copy of the p10 promoter and in pAcDZ1 they are under the control of a *Drosophila melanogaster* heat-shock promoter, which is active in Sf cells. These vectors provide for the selection of recombinant viruses which have both the desired foreign gene and the *lac*Z coding sequences inserted; the viruses may be identified by staining plaque-assays with X-gal which results in the formation of blue, polyhedrin-negative plaques.

2.4.2 p10 PROMOTER-BASED TRANSFER VECTORS

The p10 gene product is dispensible for virus particle and polyhedra formation in infected cells, although polyhedra synthesized in its absence are less stable than the normal structures (see section 2.1). There are only a few transfer vectors available for inserting foreign coding sequences under the control of the p10 promoter at the p10 locus in the virus genome. This is probably because, unlike the polyhedrin protein, synthesis of the p10 protein in infected cells does not produce a recognizable phenotype that can be used as the basis for recombinant virus selection. Recent developments (see below) have made this less of a problem and we expect to see the p10 locus used more widely as a site for inserting foreign coding sequences in the future.

A transfer vector which has been developed from the p10 promoter is pAcUW1 (Weyer *et al.*, 1990) and is shown in Figure 2.4a. This has *Bgl*II and *Hin*dIII insertion sites for foreign gene coding sequences, but lacks the facility for producing single-stranded DNA, as described for the pAcCL29 vectors above. The promoter structure of this vector is shown in Figure 2.4b; it is necessary to provide an ATG codon in the foreign DNA to be inserted into the transfer vector.

Other transfer vectors, analogous to pJV*Nhe*I and pAcDZ1 have

(a)

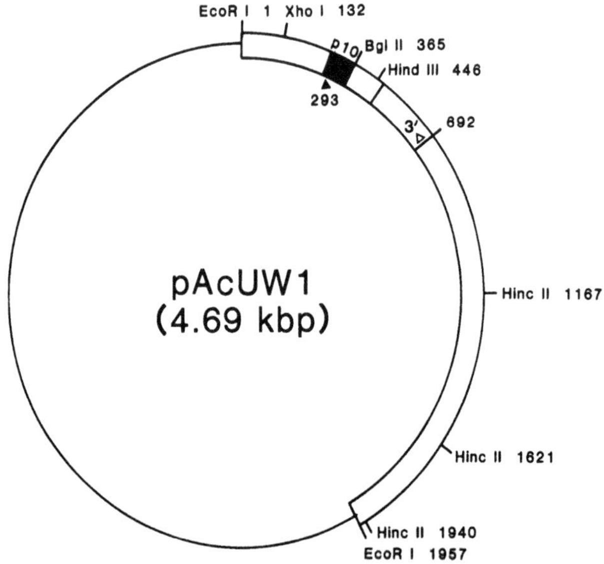

(b)

```
           140             150             160             170             180             190
CTCGAGCAAGAAAATAAAACGCCAAACGCGTTGGAGTCTTGTGTGCTATTTTACAAAGAT
Xho I
           200             210             220             230             240             250
TCAGAAATACGCATCACTTACAACAAGGGGGACTATGAAATTATGCATTTGAGGATGCCG

           260             270             280             290    T→       300             310
GGACCTTTAATTCAACCCAACACAATATATTATAGTTAAATAAGAATTATTATCAAATCA

           320             330             340             350             360             370
TTTGTATATTAATTAAAATACTATACTGTAAATTACATTTTATTTACAATCACAGATCTC
                                                                                   Bgl II
           380             390             400             410             420             430
AGAAATTCAATCCATATTGACCGGCGACATTGTTCCGGATCTTCCAGACTCACTAAAGCC

           440             450             460             470             480             490
TAAGCTGAAAAGCCAAGCTTTTGAACTCGATTCAGACGCTCGTCGTGGTAAACGCAGTTC
                  Hind III
```

Figure 2.4 (a) Restriction enzyme map of the transfer vector pAcUW1 (Weyer *et al.*, 1990). The plasmid backbone (pUC derivative) is indicated by a single line and the AcMNPV sequences by a double line. The *Bgl*II insertion site for the foreign gene sequences is at position 365 and the p10 promoter is labelled. The site of transcription initiation is shown at position 293 (solid triangle) and the probable 3' end of the mRNA is shown at position 692 (open triangle). The intervening sequences (open box) are the residual, out of frame, p10 gene coding sequences. (b) Nucleotide sequence of the p10 promoter in pAcUW1. The position of transcription initiation is double underlined and indicated by a thick, horizontal arrow. The sequences after the *Bgl*II site are derived from the residual p10 gene coding region [see part (a)].

been constructed with the *lac*Z gene inserted close to the p10 promoter, thus facilitating selection of recombinant viruses by staining with X-gal, e.g. pAcAS3 (Vlak *et al.*, 1990).

2.4.3 MULTIPLE EXPRESSION VECTORS

Baculovirus transfer vectors which can insert multiple foreign gene coding sequences into the virus genome have recently been developed. Emery and Bishop (1987) inserted the polyhedrin gene promoter and transcription termination sequences, from pAcYM1, upstream of and in the opposite orientation to, the normal polyhedrin gene. This transfer vector (pAcVC1) was used to produce a virus expressing both the lymphocytic choriomeningitis virus (LCMV) nucleoprotein (N), in addition to the polyhedrin protein. In a further experiment, the polyhedrin gene coding sequence in pAcVC1 was replaced with the LCMV glycoprotein (G) gene coding sequence, under the control of the polyhedrin promoter, to produce pAcVC2; the virus derived from pAcVC2 expressed both LCMV N and G and was the first baculovirus expression vector for two foreign coding sequences. A universal transfer vector derived from these studies (pAcVC3; V. Emery and D.H.L. Bishop, unpublished data) has *Bam*HI and *Bgl*II sites for inserting the two foreign gene coding sequences.

Although the transfer vector pAcVC3 has pioneered the expression of multiple gene products in baculovirus systems, the viruses derived from this transfer vector suffer stability problems because of the amounts of repeated sequences. The studies with the p10 promoter (Vlak *et al.*, 1988; Williams *et al.*, 1989; Weyer *et al.*, 1988, 1989), have enabled improvements to be made. A small DNA fragment (ca. 230 nucleotides), encompassing the p10 promoter, has been inserted upstream of the polyhedrin gene, at a convenient *Eco*RV site, in the opposite orientation to the polyhedrin promoter. A copy of the SV40 transcription termination signals, which had been previously demonstrated to function in baculovirus-infected cells (Possee and Howard, 1987), was also linked to this duplicated p10 promoter. A *Bgl*II insertion site was located between the promoter and transcription termination sequences; the final plasmid construction was designated pAcUW2B (see Figure 2.5; Weyer *et al.*, 1990). This vector has been used to express β-galactosidase (Weyer *et al.*, 1990) and *B. thuringiensis* delta endotoxin (Merryweather *et al.*, 1990), in addition to the polyhedrin protein. The selection of recombinant viruses was considerably simplified by the ability to restore a polyhedrin-negative mutant virus to the polyhedrin-positive phenotype (see section 2.5).

(a)

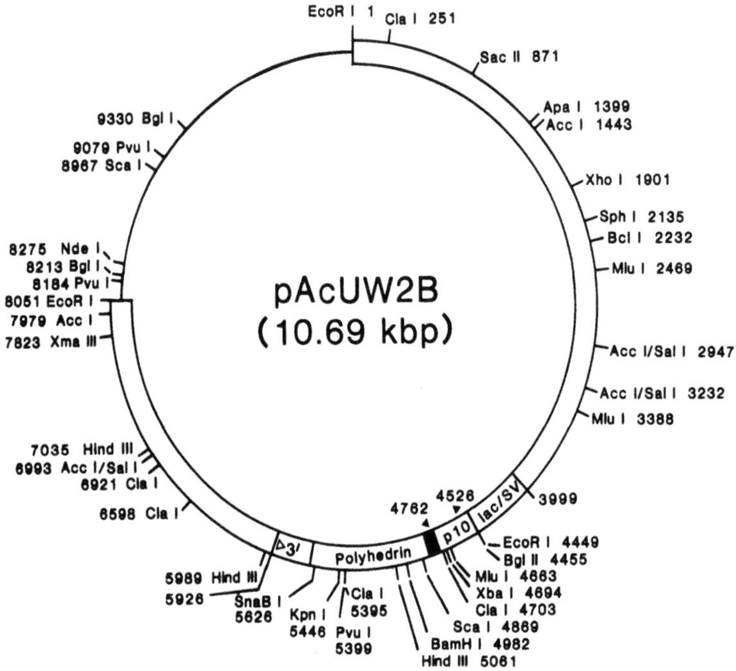

Figure 2.5 (a) Restriction enzyme map of the vector pAcUW2B (Weyer *et al.*, 1990), for the dual expression of polyhedrin and a foreign gene. The plasmid backbone (pUC derivative) is indicated by a single line and the AcMNPV and other sequences by a double line. The polyhedrin gene promoter is represented by a solid box, the polyhedrin gene coding region is labelled and the 3′ end of the mRNA is at position 5926 (open triangle). The p10 promoter is labelled; the transcription initiation site is shown at position 4526 (solid triangle) and the associated *lac*Z coding sequence (out of frame) and SV40 transcription termination signals are indicated (lacZ/SV), together with the *Bgl*II insertion site for the foreign gene. (b) Nucleotide sequence of the dual promoter region in pAcUW2B. The antisense DNA strand of the p10 promoter is shown, with the transcription initiation site double underlined and labelled with a thick horizontal arrow. The polyhedrin promoter is shown in the positive sense with the amino-terminus of the coding region illustrated.

The logical progression from pAcUW2B was to remove the polyhedrin coding sequences and thus derive a vector analogous to pAcVC3, in which two foreign gene coding sequences could be simultaneously inserted into the baculovirus genome. This vector was designated pAcUW3 (see Chapter 4, Figure 4.2; Weyer and Possee, 1991) and has been used to co-express the influenza virus haemagglutinin (HA) and neuraminidase (NA) genes in insect cells (Weyer and Possee, 1991). The structure of the promoter region of

(b)

```
          4390      4400      4410      4420      4430      4440
GCCTGCCCGGTTATTATTATTTTTGACACCAGACCAACTGGTAATGGTAGCGACCGGCGC

          4450      4460      4470      4480      4490      4500
TCAGCTGGAATTCAGATCTGTGATTGTAAATAAAATGTAATTTACAGTATAGTATTTTAA
        EcoR I Bgl II
          4510      4520      4530      4540      4550      4560
TTAATATACAAATGATTTGATAATAATTCTTATTTAACTATAATATATTGTGTTGGGTTG
                            ←T
          4570      4580      4590      4600      4610      4620
AATTAAAGGTCCCGGCATCCTCAAATGCATAATTTCATAGTCCCCCTTGTTGTAAGTGAT

          4630      4640      4650      4660      4670      4680
GCGTATTTCTGAATCTTTGTAAAATAGCACACAAGACTCCAACGCGTTTGGCGTTTTATT
                                           Mlu I
          4690      4700      4710      4720      4730      4740
TTCTTGCTCGACTCTAGAGGATCGATCCCCCGGATCTGATCATGGAGATAATTAAAATGA
            Xba I   Cla I
          4750      4760 T→    4770      4780      4790      4800
TAACCATCTCGCAAATAAATAAGTATTTTACTGTTTTCGTAACAGTTTTGTAATAAAAAA

            POLYHEDRIN
            M  P  D  Y  S  Y  R  P  T  I  G  R  T  Y  V  Y
ACCTATAAATATGCCGGATTATTCATACCGTCCCACCATCGGGCGTACCTACGTGTACGA
          4810      4820      4830      4840      4850      4860
```

pAcUW3 is also shown in Figure 4.2. The recombinant virus instability problems encountered with pAcVC3 have not been repeated with pAcUW3.

2.4.4 TRANSFER VECTORS UTILIZING OTHER BACULOVIRUS GENE PROMOTERS

The baculovirus genes active in the earlier phases of the replication cycle do not match the level of production of the polyhedrin and p10 proteins. They are, however, still potentially useful as expression vectors. A complication in their use is that most of these gene products are probably virus structural or regulatory proteins. It is unlikely, therefore, that these functions can be deleted from the virus genome in a manner analogous to the replacement of the polyhedrin and p10 genes. In consequence, it is necessary to duplicate the promoters of these genes at alternative, non-essential locations in the virus genome. One such example is the AcMNPV basic, or arginine-rich, protein which is expressed in the late phase of baculovirus replication (Wilson *et al.*, 1987). The promoter from this gene has been duplicated *in lieu* of the polyhedrin promoter in pAcRP23 to derive pAcMP1 (Hill-Perkins and Possee, 1990). This vector can be used to insert foreign genes at the polyhedrin locus and produce protein in the late phase of gene expression. The transfer vector pAcMP1 is described in Figure 2.6a and the structure of the promoter region is shown in Figure 2.6b.

2.5 SELECTION OF RECOMBINANT VIRUSES

After the co-transfection of insect cells with infectious virus DNA and a transfer vector containing the foreign gene coding sequences, the recombinant virus must be separated from the parental virus. There are now a variety of selection methods available which have

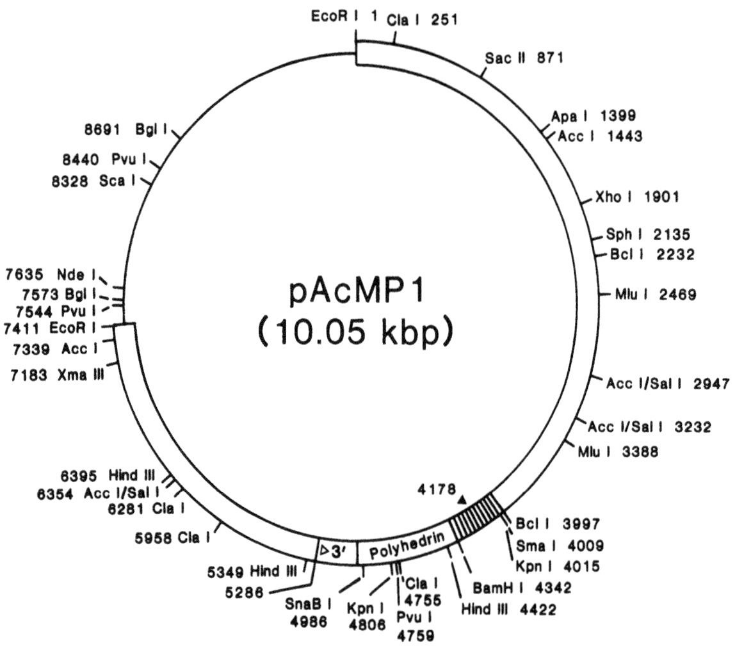

(a)

Figure 2.6 (a) Restriction enzyme map of the transfer vector pAcMP1 (Hill-Perkins and Possee, 1990). The plasmid backbone (pUC derivative) is indicated by a single line and the AcMNPV sequences by a double line. The *Bam*HI insertion site is at position 4342. The basic protein gene promoter is represented by a hatched box. The site of transcription initiation is shown at position 4178 (solid triangle) and the 3′ end of the mRNA is shown at position 5286 (open triangle). The intervening sequences are the residual, out of frame, polyhedrin gene. (b) Nucleotide sequence of the AcMNPV basic protein promoter. The position of transcription initiation is double underlined and indicated with a thick, horizontal arrow. The sequences after the *Bam*HI site are derived from the residual polyhedrin gene coding region [see part (a)].

considerably simplified what used to be a difficult stage in the production of a recombinant virus. In some of the newer methods, infectious AcMNPVDNA is replaced with DNA from a recombinant virus containing a selectable marker. These are described, together with the appropriate transfer vectors, in Table 2.1.

2.5.1 SELECTION OF A POLYHEDRIN-NEGATIVE PHENOTYPE

The original method used to select recombinant viruses exploited the fact that polyhedra are readily visible in mature plaques (3–4 days p.i.) in a standard titration of virus infectivity. In those viruses where the polyhedrin coding sequences have been replaced *via* homologous recombination, plaques lacking polyhedra are produced. With practice these virus mutants may be readily identified using two alternative methods. In the first, virus plaques without polyhedra are located by inspection using a good quality microscope. In the second method, the plaques are observed with the naked eye against a black background and illuminated with a bright light source (held at an acute angle). Polyhedrin-negative plaques appear less refractile than the polyhedrin positive plaques. The disadvantage of both of these methods is that considerable experience and practice are required to become confident of locating recombinant

```
      3910      3920      3930      3940      3950      3960
TGCGAGCAGTTGTTTGTTGTTAAAAATAACAGCCATTGTAATGAGACGCACAAACTAATA

      3970      3980      3990      4000      Kpn I     4020
TCACAAACTGGAAATGTCTATCAATATATAGTTGCTGATCAGATCCCCGGGTACCAAATT
                                   Bcl I     Sma I
      4030      4040      4050      4060      4070      4080
CCGTTTTGCGACGATGCAGAGTTTTTGAACAGGCTGCTCAAACACATAGATCCGTACCCG

      4090      4100      4110      4120      4130      4140
CTCAGTCGGATGTATTACAATGCAGCCAATACCATGTTTTACACGACTATGGAAAACTAT

      4150      4160      4170      4180      4190      4200
GCCGTGTCCAATTGCAAGTTCAACATTGAGGATTACAATAACATATTTAAGGTGATGGAA

      4210      4220      4230      4240      4250      4260
AATATTAGGAAACACAGCAACAAAAATTCAAACGACCAAGACGAGTTAAACATATATTTG

      4270      4280      4290      4300 T→   4310      4320
GGAGTTCAGTCGTCGAATGCAAAGCGTAAAAAATATTAATAAGGTAAAAATTACAGCTAC

      4330      4340      4350      4360      4370      4380
ATAAATTACACAATTTAAACGGATCCTTTCCTGGGACCCGGCAAGAACCAAAAACTCACT
                    BamH I
```

(b)

Table 2.1 List of transfer vectors and appropriate viruses to be used in co-transfection experiments

Transfer vector	*Parental virus DNA*	*Selection method for recombinant virus*	*Comments*
Polyhedrin gene locus (a) Single expression vectors			
pAcYM1[1] (polyhedrin) pEV55[2] (polyhedrin) pVL941[3] pAcCl29[4] pAcAL1–2[5] (polyhedrin) pAcMP1[6] (basic protein)	AcMNPV, or	Polyhedrin-negative plaques	Low frequency, difficult to identify
	AcMNPV.*lacZ*[8] (blue plaques), or	β-Galactosidase-negative plaques[12]	Low frequency, relatively hard to identify
	AcMNPV.*lacZ*[8] digested with *Bsu*36I/*Sau*I (linear DNA)	β-Galactosidase-negative plaques	High frequency, easy to identify
(b) dual expression vectors			
pAcUW3[7] (polyhedrin and p10)	AcMNPV, or	Polyhedrin-negative plaques[12]	Low frequency, difficult to identify
	AcRP6.SC[10] digested with *Bsu*36I/*Sau*I (linear, polhedrin-negative DNA)	Random plaques for screening by dot-bolt or PCR	High frequency, relatively easy to identify
pAcUW2B[9] (p10 plus normal polyhedrin gene	AcRP6.SC[10] digested with *Bsu*36I/*Sau*I (linear, polyhedrin-negative DNA), or	Polyhedrin-positive plaques	High frequency, easy to identify
	Any polyhedrin-negative AcMNPV	Polyhedrin-positive plaques	Low frequency, but easy to identify
p10 gene locus			
pAcUW1[9] (p10)	AcUW1.*lacZ*[9,11] (blue plaques), or	β-Galactosidase-negative plaques[12]	Low frequency, relatively hard to identify
	AcUW1.*lacZ*[9,11] digested with *Bsu36I*/*Sau*I (linear DNA)	β-Galactosidase-negative plaques	High frequency, easy to identify

[1] Matsuura *et al.* (1987). [2] Miller *et al.* (1986). [3] Luckow and Summers (1989). [4] Livingstone and Jones (1989). [5] See Chapter 2. [6] Hill-Perkins and Possee (1990). [7] Weyer and Possee (1991). [8] AcMNPV.*lacZ* contains the *lacZ* coding sequences, in place of polyhedrin, at the polyhedrin gene locus. [9] Weyer *et al.* (1990). [10] Kitts *et al.* (1990). [11] AcUW1.*lacZ* contains the *lacZ* coding sequences, in place of p10, at the p10 gene locus. [12] Dot-blot hybridization/PCR may also be used.

plaques. This method is also dependent on producing good quality plaque-assays. Although this may be a trivial matter for virologists, it is more difficult for the novice and can cause many weeks of frustration. It is our experience that guidance from someone well versed in the preparation of recombinant viruses using selection of the polyhedrin-negative phenotype is invaluable for the newcomer to the expression system. The frequency of recombination, estimated at about 1% when scoring total numbers of recombinant viruses using *lac*Z as a marker (Kitts *et al.*, 1990), may be considerably lower when screening for the loss of polyhedrin phenotype; presumably because such plaques are frequently obscured by the wild-type virus.

2.5.2 SELECTION OF β-GALACTOSIDASE-NEGATIVE VIRUSES

The *E. coli* β-galactosidase (*lac*Z) gene confers a blue phenotype on the recombinant virus when expressed in insect cells in the presence of X-gal (Pennock *et al.*, 1984; Possee and Howard, 1987; Summers and Smith, 1987). If infectious DNA from such a virus (e.g. AcMNPV.*lac*Z; Possee and Howard, 1987) is used in place of the wild-type AcMNPV DNA during a co-transfection, then replacement of the *lac*Z gene derives a colourless phenotype. Originally this appeared an attractive method for selecting recombinant virus. Unfortunately, it is our experience that it is almost as difficult to locate these 'colourless' plaques, as it is the polyhedrin-negative variety. The major problem appears to be the masking of putative recombinants with blue, parental AcMNPV.*lac*Z plaques.

2.5.3 RECOMBINANT VIRUS SELECTION USING DOT-BLOT HYBRIDIZATION

A useful alternative to the above methods is to identify recombinant viruses using dot-blot hybridization. This may done in two ways. In the first, putative recombinant plaques are identified with one of the above methods and used to establish an infection in cells growing in a 96-well microtitre plate (Summers and Smith, 1987). In the second, dilutions of the progeny virus from the original co-transfection are used to infect cells in microtitre plates (Fung *et al.*, 1988). The DNA is then extracted from cells in both methods and hybridized with a radiolabelled copy of the foreign gene coding sequences. The virus derived from the positive wells is then subjected to further rounds of plaque purification and/or dot-blot hybridization to yield the homogeneous recombinant. The method works well, because it relies less on subjective examination of virus plaques and more on objective hybridization tests. It is, however, relatively time consuming.

2.5.4 SCREENING FOR A POSITIVE PHENOTYPE

The traditional methods used to screen for recombinant viruses rely on identifying the loss of a particular phenotype (e.g. polyhedra or β-galactosidase), as described in sections 2.5.1 and 2.5.2. This approach may be reversed so that the infectious virus DNA has the 'negative-phenotype' and the transfer vector confers the 'positive-phenotype' on the recombinant virus.

Vialard *et al.* (1989) described a transfer vector which has the *lacZ* gene, under the control of a p10 promoter, inserted upstream from the polyhedrin promoter and foreign coding sequences (see section 2.2). When used in co-transfections with wild-type AcMNPV DNA, the recombinant progeny may be easily identified as blue, polyhedrin-negative plaques. The system has the added advantage that the plaque-assays need not be of such a high standard as when screening only for the polyhedrin-negative phenotype. The disadvantage of the method is that the recombinant virus will produce β-galactosidase, which will always be present as a contaminant in protein preparations. Furthermore, the transfer vector containing this arrangement of selectable marker and foreign coding sequences becomes rather large and potentially unstable. A second method which uses an element of positive-selection was described by Weyer *et al.* (1990). A transfer vector was constructed with a copy of the p10 promoter and a RE cloning site inserted upstream of the intact polyhedrin gene. When used in co-transfections with DNA from a polyhedrin-negative mutant virus, the resulting polyhedrin-positive recombinants were relatively easy to identify. The disadvantage of this method, as with the use of β-galactosidase, is that polyhedrin protein will be a contaminant of cell extracts; although if the recombinant protein is located in the cytoplasm the nuclei containing the polyhedra may be easily removed using centrifugation, after lysing the cells with non-ionic detergents. A particular advantage of this method, however, is that the polyhedrin-positive recombinants may be used efficiently to infect insect larvae. If the foreign protein to be expressed is potentially insecticidal then this is the best way to test its activity in insects; if a polyhedrin-negative virus is used, the insects must be infected by injection directly into the haemolymph.

2.5.5 ENHANCING THE NUMBERS OF RECOMBINANT VIRUSES

None of the methods described above provide the ideal solution to the problem of isolating recombinant viruses. They fail to address the question of the low frequency of recombination in a co-transfection and merely provide alternative procedures for

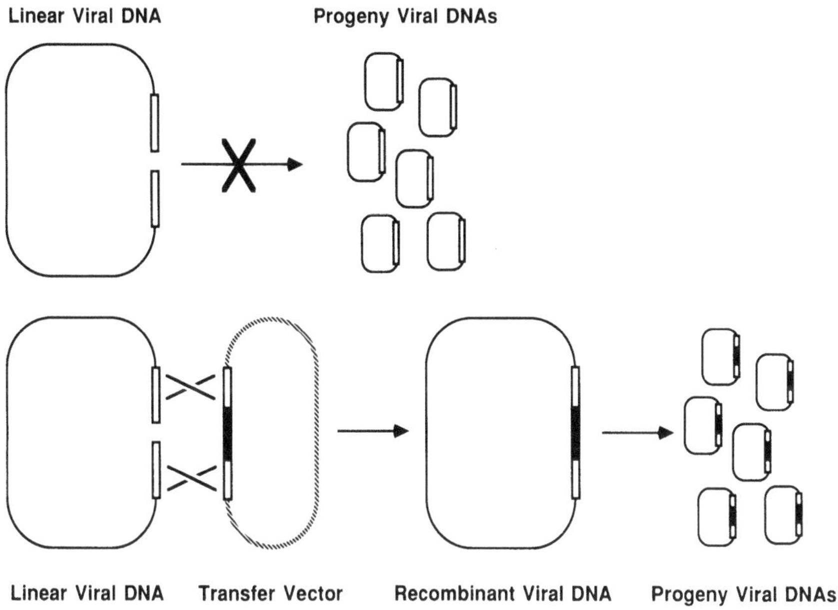

Figure 2.7 Rescue of linear virus DNA by recombination with a transfer vector. Upper panel: linear DNA cannot replicate because the replication apparatus of AcMNPV is designed to work on the native virus DNA which is circular. Lower panel: the circularity of the virus DNA can be restored by recombination with a transfer vector carrying DNA homologous to the virus sequences on either side of the break. A double cross-over generates a recombinant virus DNA molecule which, being circular, is competent for replication. Reproduced from Kitts *et al.* (1990) with permission.

identifying the expression vector. A radically different approach has recently been described by Kitts *et al.* (1990), which does provide a method for enhancing the recovery of recombinant viruses. This uses linearized baculovirus DNA, in combination with standard transfer vectors such as pAcYM1, pEV55, pAcRP23 and pVL941, which results in the production of 30–40% recombinant viruses in the first round of plaque purification.

The method was based on the observation that, in yeast and mammalian cells, DNA molecules which have double-stranded breaks are highly recombinogenic (Orr-Weaver *et al.*, 1981; Bollag *et al.*, 1989). This phenomenon has been expoited to promote recombination with homologous sequences in these hosts by linearizing the DNA prior to introducing it to the cell (Orr Weaver *et al.*, 1983; Bollag *et al.*, 1989). In insect cells, linear fragments of AcMNPV DNA, produced by restriction enzymes, have been used to map mutations in the virus genome by marker rescue (Miller, 1981; Gordon and Carstens, 1984), but linear baculovirus DNA is generally regarded as non-infectious (Kelly and Wang, 1981). These results were used as the basis for designing a new recombinant virus selection system. It was proposed that a linearized baculovirus genome might be non-infectious, but still be able to recombine with homologous sequences introduced into the host cell. If the homologous sequences spanned the two ends of the linearized baculovirus DNA, then the genome would be recircularized and restored to full infectivity. The principle of the method is outlined in Figure 2.7.

This hypothesis was tested when a RE (*Bsu*36I/*Mst*II/*Sau*I) was identified which did not digest AcMNPV DNA. A synthetic oligonucleotide, containing this site was introduced into the transfer vector pAcRP6 (Matsuura *et al.*, 1987), at the *Bam*HI insertion site for foreign coding sequences, downstream of the polyhedrin promoter. The modified transfer vector (pAcRP6.SC; SC is for **S**ingle **C**ut) was used to derive a polyhedrin-negative virus, designated AcRP6.SC or AcMNPV.SC (this book). When linearized (*Bsu*36I-digested) AcMNPV.SC DNA, instead of circular AcMNPV DNA, was used to transfect Sf cells, the yield of progeny virus was reduced by 93%. This confirmed that baculovirus DNA should be circular for full infectivity. In subsequent tests, the linearized virus DNA was co-transfected with pAcRP23.*lac*Z (a transfer vector with the *E. coli lac*Z coding sequences inserted under the control of the polyhedrin promoter). The virus derived from this co-transfection was titrated in a plaque-assay, and recombinant virus plaques detected by staining blue in the presence of X-gal. These results showed that between 16 and 39% of the plaques produced were from recombinant viruses.

The use of linearized virus DNA in co-transfections represented a considerable improvement over the previous methods for selecting recombinant virus. The system has been refined further by the discovery that the *Bsu*36I (*Mst*II/*Sau*I) RE site is also present in the bacterial *lac*Z gene coding sequences. A virus (AcMNPV.*lac*Z; Possee and Howard, 1987) with this sequence inserted *in lieu* of the polyhedrin gene can, therefore, be linearized in the same way as AcMNPV.SC, prior to performing co-transfections with the transfer vectors. These experiments showed that after titrating the progeny virus in a plaque-assay, many colourless-plaques were observed after staining with X-gal. Subsequent analysis of these plaques revealed that 40–80% of the colourless-plaques had the appropriate foreign coding sequences inserted. This method, therefore, provides an extremely powerful tool in the derivation of recombinant baculoviruses and is now the first choice in many laboratories. Protocols for the preparation and use of linearized virus DNA in co-transfections will be described in Chapter 7.

Processing of foreign proteins synthesized using baculovirus vectors in insect cells 3

3.1 INTRODUCTION

The widespread use and success of the baculovirus system makes it impractical to provide an exhaustive survey of all the foreign proteins synthesized in insect cells. Instead, it is our aim to consider the co- and post-translational processing events that have been demonstrated to occur in insect cells and to illustrate these with selected examples.

The insect cell appears capable of undertaking many of the processing events that are required for the formation of biologically active, heterologous proteins. The following processing events will be discussed: glycosylation; phosphorylation; fatty acid acylation; amidation; proteolytic processing, including signal-peptide cleavage; cellular targeting and secretion; tertiary and quaternary structure formation. Post-transcriptional processing will also be mentioned very briefly, although at the present time there is little data on this topic. Several recent review articles will provide the reader with more comprehensive surveys of genes that have been expressed using baculovirus vectors (Fraser, 1989; Maeda, 1989a; Miller, 1988; Atkinson *et al.*, 1990a; Bishop and Possee, 1990; Possee *et al.*, 1990; Vlak and Keus, 1990; Luckow, 1990).

The amounts of protein synthesized by recombinant baculoviruses in insect cells vary from a yield equivalent to that of polyhedrin (approximately 500 mg/litre cells), to relatively low yields of less than 1 mg/litre cells. Unfortunately, it is not yet possible to predict with any confidence how well a particular protein will be synthesized,

Table 3.1 Examples of glycoproteins expressed using baculovirus vectors

Protein	*Species/virus*	*Membrane-targeted (MT) Secreted (S)*	*Reference*
Virus examples			
env-gp85	Avian leukaemia virus	MT[1]	Noteborn *et al.* (1990)
Spike gp	Bovine corona virus	S	Yoo *et al.* (1991)
Haemagglutinin	Fowl-plague virus	MT[1,2]	Kuroda *et al.* (1986)
Surface antigen	Hepatitis B virus	S[1,2]	Kang *et al.* (1987) Takehara *et al.* (1988) Lanford *et al.* (1989)
Glycoprotein D	Herpes simplex virus (type 1)	MT[1,2]	Krishna *et al.* (1989)
gp160	Human immuno-deficiency virus	MT[1,2]	Rusche *et al.* (1987) Hu *et al.* (1987)
Haemagglutinin	Human influenza virus	MT	Possee (1986) Kuroda *et al.* (1989) Kuroda *et al.* (1990)
Fusion glycoprotein	Human parainfluenza virus	MT[1,2]	Ray *et al.* (1989)
F glycoprotein	Human respiratory syncytial virus	MT[1,2]	Wathen *et al.* (1989)
Haemagglutinin	Japanese encephalitis virus	MT	Matsuura *et al.* (1989)
Membrane fusion and env glycoprotein	Measles virus	MT	Vialard *et al.* (1990)
Peplomer gp (E2)	Murine coronavirus (JHM)	MT[1]	Yoden *et al.* (1989)
HN proteins	Parainfluenza virus (type 3)	MT	Van Wyke Coelingh *et al.* (1987)
G protein	Rabies virus	MT[1,2]	Préhaud *et al.* (1989)
Glycoprotein	Vesicular stomatitis virus	MT	Bailey *et al.* (1989)
E and NS1	Yellow fever virus	–	Desprès *et al.* (1991)
Non-virus examples			
Diuretic hormone	*M. sexta* (tobacco hornworm)	S[1]	Maeda (1989b)
Juvenile hormone esterase	*Heliothis virescens* (tobacco budworm)	S[1]	Hammock *et al.* (1990)
Acid β-gluco-sidase	Human	MT	Grabowski *et al.* (1989)
β-adrenergic receptor	Human	–	George *et al.* (1989)
CD4 receptor	Human	MT[1]	Webb *et al.* (1989)
EGF receptor	Human	MT[1]	Greenfield *et al.* (1988)
Glucocere-brosidase	Human	S[1]	Bergh *et al.* (1990) Martin *et al.* (1988)
Haptoglobin	Human	S[1]	Heinderyckx *et al.* (1989)

Protein	*Species/virus*	*Membrane -targeted (MT) Secreted (S)*	*Reference*
Immunoglobulin Heavy chain (γ-1)	Human	S[1]	Hasemann and Capra (1990)
Immunoglobulin Light chain (91A3)	Human	S[1]	Hasemann and Capra (1990)
Insulin receptor	Human	MT	Herrera *et al.* (1988) Paul *et al.* (1990)
β-interferon	Human	S[1]	Smith *et al.* (1983)
Myelin-associated glycoprotein	Human	S[1]	Johnson *et al.* (1989)
Plasminogen	Human	S[1]	Davidson *et al.* (1990) Whitefleet-Smith *et al.* (1989)
Chimeric plasminogen activators (PA)	Human	S	Devlin *et al.* (1989)
Poliovirus receptor	Human	–	Kaplan *et al.* (1990)
Tissue-type PA	Human	S[1]	Jarvis and Summers (1989)
Transferrin receptor	Human	MT[1]	Domingo and Trowbridge (1988)
Urokinase-type PA	Human	S	King *et al.* (1991b)
$GABA_A$ receptor	Bovine	MT	Atkinson *et al.* (1991)
Phaseolin	*Phaselolus vulgaris* (French bean)	S[1]	Bustos *et al.* (1988)
Ricin B-chain	*Ricinus communis* (Castor bean)	S[1]	Piatak *et al.* (1988)

[1] Antigenic.
[2] Elicited neutralizing antibodies.

although it appears that yields of membrane-associated glycoproteins tend to be on the low side. We hope that the summary in this chapter will provide some useful guidelines as to what may be expected when expressing a new foreign gene.

3.2 GLYCOSYLATION

Many biologically active glycoproteins have been synthesized in insect cells and some examples are listed in Table 3.1. The most common and thoroughly investigated type of glycosylation is the N-linked process that occurs in the endoplasmic reticulum of both mammalian and insect cells. N-linked glycosylation is initiated by the addition of an oligosaccharide to an asparagine residue (consensus recognition site, Asp-X-Ser/Thr) and is thought to be mediated via a phospholipid carrier (reviewed by Kornfield and Kornfield, 1985; Parekh *et al.*, 1989).

In mammalian and insect cells, N-linked glycosylation may be inhibited by treating cells with tunicamycin. Comparison of mammalian cell- and insect cell-derived glycoproteins, however, has revealed several differences in the nature of the added oligosaccharide side-chains. Mammalian cells extensively trim and then modify the core oligosaccharide in terminal glycosylation events. These events usually result in the addition of fucose, glucosamine-galactose and sialic acid residues to form complex, branched oligosaccharides. Insect cells appear to lack, or have only very low levels, of the necessary fucose, galactose and sialic acid transferases. They appear to lack the ability to process the core oligosaccharide (Butters and Hughes, 1981; Butters *et al.*, 1981). Furthermore, in insect cells the core oligosaccharide has been shown to contain a high proportion of mannose residues. Recombinant glycoproteins synthesized in insect cells may, therefore, be readily detected by radiolabelling with [^{3}H] or [^{14}C]mannose, or by binding to mannose-specific lectins such as concanavalin A. From the few detailed studies performed on recombinant glycoproteins produced by baculoviruses in insect cells, it would appear that about 50% of the core oligosaccharides are trimmed to GlcNAc2Man3, with the remainder having a variable number of mannose residues (Kuroda *et al.*, 1990).

One consequence of these differences in glycosylation is that recombinant proteins in insect cells have a smaller apparent molecular weight, after analysis by PAGE, than their authentic counterparts in the normal host cell. For example, the avian leukaemia virus (ALV) envelope glycoprotein (env gp) has an apparent molecular weight of 85 kDa, whereas the insect cell-expressed protein has molecular weights ranging from 45 to 65 kDa (Noteborn *et al.*, 1990).

Baculovirus-expressed recombinant proteins are sensitive to endo H, endo F and N-glycanase, which remove immature high-mannose type oligosaccharides. The treatment of insect cell-derived recombinant proteins with these enzymes (or treatment of infected-cells with tunicamycin) results in proteins that have the same molecular weight as similarly treated proteins derived from the normal host cell. For example, when the ALV env gp85 was synthesized in the presence of tunicamycin, both the avian and insect cell-derived protein had a molecular weight of 45 kDa (Noteborn *et al.*, 1990). These results have been complemented by a study on the role played by glycosylation in the secretion of tissue plasminogen activator (t-PA) from insect cells (Jarvis *et al.*, 1990b). In this study, treatment of recombinant virus-infected cells with tunicamycin inhibited secretion of t-PA. However, treatment of cells with inhibitors of N-linked oligosaccharide processing (castanospermine or *N*-methyldeoxynojirimycin) did not inhibit t-PA secretion. These results suggested that addition of N-linked oligosaccharides, but not their subsequent processing, was necessary for t-PA secretion from insect cells.

The first evidence that limited processing of the core oligosaccharides of recombinant glycoproteins can occur in insect cells has been provided by Davidson *et al.* (1990). In this study, recombinant human plasminogen was synthesized in Sf9 cells using a serum-free medium (EX-CELL 400, see Chapter 5). Analysis of the asparagine-linked oligosaccharides revealed that 60% of the side-chains were of the high-mannose type, as described above. The remaining 40% were of a more complex type consisting primarily of (GlcNAc$_2$-Man-(Man-GlcNAc-Gal-Sia)$_2$), indicating that mannosidase, galactosyl-hexaminidasyl- and sialyl-transferases were present or could be induced in cultured insect cells.

Another report has demonstrated that insect cells are able to perform O-linked glycosylation (Thomsen *et al.*, 1990). In this study pseudorabies gp50, which has no consensus N-linked glycosylation sites, was shown to be glycosylated by radiolabelling infected cells with [^{14}C]glucosamine. Further analyses demonstrated the nature of the oligosaccharide side-chains to be mainly GalNAc, with lower amounts of Galβ1-3GalNAc; no sialic acid residues were detected. Measurement of the relevant glycosyltransferases revealed that Sf9 cells contained similar levels of UDP-GalNAc, N-acetylgalactosaminyltransferases as Vero and CHO cells, but lower levels of UDP-Gal:N-acetylgalactosamine-β1-3galactosyltransferase activity.

Where assays of biological activity have been carried out, glycoproteins synthesized in insect cells have been shown to have similar, if not identical, activities to their authentic counterparts. For example, influenza virus, fowl-plague virus (FPV), and parainfluenza virus (type 3) haemagglutinins (HA) were able to haemagglutinate red blood cells (Kuroda *et al.*, 1986; Possee, 1986; Van Wyke Coelingh *et al.*, 1987). Several other virus glycoproteins have been expressed in insect cells and all have either been membrane targeted or secreted, as appropriate; some examples of these have been included in Table 3.1. In each of these examples the glycoproteins synthesized were antigenic, giving rise to high-titre antibody preparations after injection into animals. However, not all the antibodies raised were neutralizing or gave protection against challenge by the live virus; whether this may be attributed to differences in glycosylation is debatable.

Both human t-PA (Jarvis and Summers, 1989) and urinary-type plasminogen activator (u-PA; King *et al.*, 1991b) have been secreted in active forms. The baculovirus expressed u-PA was shown to dissolve fibrin clots in both fibrin-well assays and by using zymography. Injection of juvenile hormone esterase, secreted from insect cells infected with a recombinant baculovirus, into *Manduca sexta* larvae caused the caterpillars to turn black (melanize) at the next moult (Hammock *et al.*, 1990). Melanization is a common assay for anti-juvenile hormone activity.

Many glycoproteins have been correctly targeted and inserted into the insect cell plasma membrane. These include the three HA proteins described above and also the human epidermal growth factor (EGF) receptor (Greenfield *et al.*, 1988), the human insulin receptor (Herrera *et al.*, 1988; Paul *et al.*, 1990), the human CD4 receptor (Webb *et al.*, 1989) and the chick nicotinic acetylcholine receptor α-subunit (Atkinson *et al.*, 1990b). This topic is discussed in more detail in section 3.5.

In summary, there are differences in glycosylation between insect and other cell types. Whether this has a bearing on the biological activity or antigenicity of any given glycoprotein will probably have to be tested empirically for each example. There are some concerns that differences in glycosylation may affect the usefulness of potential therapeutic proteins made in insect cells. For example, concern has been expressed about differences in antigenicity (Anicetti *et al.*, 1989; Marino, 1989) and in clearance rates from plasma, since one study has shown that glycoproteins with oligosaccharides terminating in mannose, GlcNAc or galactose are rapidly cleared from plasma (Stahl, 1990). There is some preliminary evidence that different types of insect cells (derived from different species or tissue types) may have alternative glycosylation pathways to those described above for *S. frugiperda* cells (Klenk, H. D., personal communication) and these may prove to be useful for glycoprotein synthesis. In the future, it may also prove possible to engineer insect cell lines to encode mammalian oligosaccharide transferases and thus complete the trimming and processing of the core oligosaccharide.

3.3 PHOSPHORYLATION, ACYLATION AND AMIDATION

Phosphorylation of a number of recombinant proteins expressed in insect cells have been reported and examples of these are listed in Table 3.2. As will be described in section 3.5, many of the recombinant proteins that have been shown to be targeted to the nucleus are phosphorylated. The combined data from these reports

Table 3.2 Examples of foreign proteins that have been phosphorylated, acylated or amidated in insect cells

Protein	*Species/virus*	*Phosphorylated (P) Acylated (Ac)*[1] *Amidated (Am)*	*Reference*
Virus examples			
E1A	Adenovirus	P	Patel *et al.* (1988)
E2 protein	Bovine papillomavirus (type 1)	P	McBride *et al.* (1989)
Core antigen	Hepatitis B virus	P	Lanford and Notvall (1990)
Surface antigen	Hepatitis B virus	Ac-M	Lanford *et al.* (1989)
$p17^{gag}$	HIV	Ac-M	Overton *et al.* (1989)
$p24^{gag}$	HIV	P	Overton *et al.* (1989)
$p40^{x}$ trans-activator	Human T-cell leukaemia virus (HTLV- I)	P	Jeang *et al.* (1987a) Nyunoya *et al.* (1988)
Nucleoprotein	Rabies virus	P	Préhaud *et al.* (1990)
Large T antigen	SV40 virus	P, Ac-P	Lanford (1988) Murphy *et al.* (1988)
gag precursor	Simian immuno-deficiency virus (SIV)	Ac-M	Delchambre *et al.* (1989)
Non-virus examples			
Krüppel	*Drosophila*	P	Ollo and Maniatis (1987)
Diuretic hormone	*M. sexta*	Am	Maeda (1989b)
EGF receptor	Human	P	Greenfield *et al.* (1988)
Insulin receptor	Human	P	Herrera *et al.* (1988) Paul *et al.* (1990)
c-*myc* proto oncogene	Human	P	Miyamoto *et al.* (1985)
$pp60^{c\text{-}src}$ oncogene	Human	P,Ac-M	Piwnica-Worms *et al.* (1990)
P^{210} BCR-ABL oncogene	Human	P	Pendergast *et al.* (1989)
Terminal transferase	Human	P	Chang *et al.* (1988)
Transferrin receptor	Human	Ac-P	Domingo and Trowbridge (1988)
Protein kinase C-γ	Bovine	P	Patel *et al.* (1988, 1989)
p53	Murine	P	O'Reilly and Miller (1988)
Tyrosine hydroxylase	Rat	P	Fitzpatrick *et al.* (1989)
$pp60^{v\text{-}src}$ (rsk- α/β)	*Xenopus*	P	Vik *et al.* (1990)
Transposon Ac	*Zea mays* (corn)	P	Hauser *et al.* (1988)

[1] Ac-M = myristylation. Ac-P = palmitylation.

suggest that phosphorylation is carried out by an endogenous protein kinase, with the major phosphoamino acid being phosphoserine (Ellis *et al.*, 1988; Nyunoya *et al.*, 1988). The fidelity of phosphorylation in insect cells has been demonstrated by the autophosphorylation of a number of recombinant proteins, including the human EGF receptor (Greenfield *et al.*, 1988), the human insulin receptor (Herrera *et al.*, 1988; Paul *et al.*, 1990) and the P^{210} BCR-ABL oncogene product, associated with chronic myelogenous leukaemia (Pendergast *et al.*, 1989). One recent study has shown that the $pp60^{c\text{-}src}$ oncogene product is phosphorylated at tyr 416, and that this activated kinase can then phosphorylate co-expressed polyoma virus middle T antigen (Piwnica-Worms *et al.*, 1990). Overton *et al.* (1989) have demonstrated that human immunodeficiency virus (HIV) $p55^{gag}$ is processed to give p24 and p17 when co- expressed with HIV protease. The p24 gene product is subsequently phosphorylated and the p17 gene product is acylated.

At least two types of fatty acid acylation have been demonstrated in insect cells: palmitylation and myristylation. The SV40 large T antigen has been labelled with [^{3}H] palmitic acid, a normal processing requirement for the membrane-targeting of this viral protein (Lanford, 1988). The hepatitis B virus surface antigen (Lanford *et al.*, 1989) and the HIV gag p17 gene product (Overton *et al.*, 1989) are acylated with myristic acid at the amino-terminus.

To date no amidation activity has been demonstrated in the processing of mammalian or virus proteins in insect cells *in vitro*, although it is well documented that insect larvae produce hormones that are amidated, for example, the diuretic hormone of *M. sexta* (Kataoka *et al.*, 1989). When this diuretic hormone was synthesized using a baculovirus vector, it was found to be amidated at the C-terminus (Maeda, 1989b). One study by Lebacq-Verheyden *et al.* (1988) has shown that, although the gastrin-releasing peptide precursor is cleaved correctly in insect cells, the processed peptide is not amidated and consequently is not fully active.

3.4 PROTEOLYTIC PROCESSING

Insect cell proteolytic enzymes appear capable of recognizing and accurately cleaving a number of recombinant proteins. This section will focus on proteolytic processing events excluding cleavage of signal peptides, which will be discussed in section 3.5. Trypsin-like endoprotease and carboxypeptidase N activities in insect cells have been inferred from the correct maturation of the FPV HA (Kuroda *et al.*, 1986, 1989). In these studies haemolytic fusion activity of

FPV HA, which is dependent upon cleavage of HA into HA1 and HA2 subunits, was demonstrated. In another example, the correct processing of human respiratory syncytial virus (RSV) F glycoprotein was demonstrated. The precursor F0 (68 kDa) was cleaved to give two subunits, F1 (48 kDa) and F2 (20 kDa), which were held together by disulphide bonds (Wathen *et al.*, 1989).

Two other studies, however, did not detect endoproteolytic cleavages. The HA of the A/PR/8/34 strain of influenza virus was not processed to HA1 and HA2 (Possee, 1986) and the HIV gp160 was not cleaved into gp120 and gp41 (Rusche *et al.*, 1987). In a more recent study, HIV p55gag was shown to be processed correctly, to give p17 and p24, only when co-expressed with the HIV protease gene (Overton *et al.*, 1989).

It would therefore appear that some viral proteins are correctly cleaved in insect cells, whereas, others require concomitant expression of the necessary virus-encoded proteases. There are several examples of virus polyproteins, encoding viral proteases, that are correctly processed in insect cells. These include Sindbis virus (Oker-Blom and Summers, 1989), poliovirus type 3 (Urakawa *et al.*, 1989) and foot-and-mouth-disease virus (FMDV) (Roosien *et al.*, 1990). The poliovirus example is perhaps the most striking. In this study, the 6.6 kb coding region of poliovirus type 3 was expressed in insect cells and produced non-infectious, empty poliovirus capsids (Urakawa *et al.*, 1989). The capsids had the configuration VP0, VP1 and VP3. Thus the insect cells had permitted all the complex processing events in the maturation of the poliovirus virion, with the exception of the final cleavage of VP0 to VP2 and VP4, which is auto-catalytic upon encapsidation of the virus RNA genome.

3.5 CELLULAR TARGETING AND SECRETION

Proteins synthesized by baculovirus vectors appear to be translocated in the insect cell as would be expected for the normal protein in its own host cell. For example, most signal peptides are recognized and cleaved in the endoplasmic reticulum and the resulting protein is targeted to the membrane or secreted, as appropriate. Amino-terminal sequence analysis has confirmed the correct signal peptide cleavage for human α-interferon (Maeda *et al.*, 1985), human interleukin-2 (IL2; Smith *et al.*, 1985), juvenile hormone esterase (Hammock *et al.*, 1990) and the ALV env gp85 (Noteborn *et al.*, 1990). In one detailed study, Lebacq-Verheyden *et al.* (1988) demonstrated that human gastrin-releasing peptide precursor (GRP1-125) was correctly cleaved to give an active peptide, GRP1-27 and a C-terminal portion

GRP31-125. However, a number of other novel cleavage forms were also identified that were not found when GRP was processed in a mammalian lung cancer cell line.

Membrane targeting also depends on the presence of a hydrophobic, membrane anchor sequence. Deletion of this region may enable the protein to be secreted. For example, the *Torpedo* nicotinic acetylcholine receptor α-subunit is secreted from insect cells if the membrane anchor region is deleted (Atkinson, A.E., Akhtar, R., Barkas, T. and King, L.A., unpublished data). Secretion of proteins has particular advantages in the purification of gene products. As with other expression systems, however, it is probably not always sufficient simply to add a signal peptide to a protein that is not normally secreted and expect it to translocate to the culture medium. The nature of the foreign protein may preclude its passage through membranes.

Examples of recombinant glycoproteins that have been efficiently secreted or membrane-targeted are listed in Table 3.1. A number of other recombinant proteins that appear not to have been glycosylated have also been efficiently secreted from insect cells, including: ricin (from *Ricinus communis*, castor bean; Piatak *et al.*, 1988); the human immune activation protein, act-2 (Lipes *et al.*, 1988); the extracellular domain of nerve growth factor receptor (Vissavajjhala and Ross, 1990) and platelet-derived growth factor (Giese *et al.*, 1989).

Baculovirus-expressed proteins may also be found in the cytoplasm or nucleus. Many of the proteins targeted to the nucleus are phosphorylated (see section 3.3), for example, the c-*myc* gene product (Miyamoto *et al.*, 1985), the *Drosophila* Krüppel gene product (Ollo and Maniatis, 1987) and the SV40 and polyoma virus large T antigens (Rice *et al.*, 1987; Murphy *et al.*, 1988). Others are located in the nucleus without phosphorylation, for example, Dengue virus core protein (Makino *et al.*, 1989). It still remains to be conclusively demonstrated that insect cells use the same nuclear transfer signals as mammalian cells.

Proteins located in the cytoplasm may form insoluble inclusion bodies, such as the lymphocytic choriomeningitis virus N protein (Matsuura *et al.*, 1987) and the Lassa fever virus nucleocapsid protein (Barber *et al.*, 1990), or may be completely soluble such as the human glucocorticoid receptor (Srinivasan and Thompson, 1990). Other recombinant proteins may be only partly soluble, for example, the catalytic subunit of protein phosphatase 1, a key enzyme in the regulation of many cellular functions (Berndt and Cohen, 1990). This protein was synthesized to high levels (25% total cell protein) in the cytoplasm where approximately 5% was found as a soluble, active species and the remaining 95% as an insoluble, inactive species. The insoluble species could be dissolved in 6 M guanidinium chloride to give a fully active protein.

3.6 TERTIARY AND QUATERNARY STRUCTURE FORMATION

Perhaps one of the most useful features of the baculovirus system is the ability of the recombinant proteins to form tertiary and complex quaternary structures. In particular, the multiple expression capabilities of the system allow two or more proteins to be synthesized simultaneously and thus any interactions forming between the proteins can be analysed (see Chapter 2). The formation of a complex between two of the three influenza virus polymerase proteins (St Angelo *et al.*, 1987), between the SV40 large T antigen and human p53 proteins (O'Reilly and Miller, 1988), and between the HIV gp120 and the soluble domains of the CD4 (SCD4) receptor (Morikawa *et al.*, 1990) have been demonstrated. Co-infection of insect cells with three recombinant viruses expressing the UL5, UL8 and UL52 genes of herpes simplex virus (HSV) type 1 produced a functional helicase–primase complex (Dodson *et al.*, 1989). This complex consisted of three polypeptide chains with a combined molecular weight of 270 kDa and was shown to have authentic DNA-dependent ATPase and GTPase, DNA helicase and DNA primase activities.

The formation of discrete sub-viral particles has been observed for the hepatitis B virus (HBV) surface (s) and core (c) antigens (Kang *et al.*, 1987; Takehara *et al.*, 1988; Lanford *et al.*, 1989). Kang *et al.* (1987) demonstrated the accumulation of 22 nm particles in the culture medium of insect cells expressing the HBVs antigen. These particles appeared to be identical to the 22 nm particles found in the plasma of patients with chronic hepatitis. Co-expression of the HBV s and c antigens gave rise to 27 nm particles (Takehara *et al.*, 1988).

In insect cells infected with recombinant viruses expressing the bluetongue virus (BTV) 10 NS1 gene product, tubule structures were formed similar to those observed in mammalian cells infected with BTV (Urakawa and Roy, 1988). Further studies on the morphogenesis of bluetongue virus have been carried out by making use of baculovirus multiple expression vectors. Core-like structures were observed when the two major BTV core proteins were co-expressed (French and Roy, 1990) and a double-shelled, virus-like particle was assembled when four BTV proteins were expressed simultaneously (French *et al.*, 1990). In this example, multiple expression was achieved by the co-infection of cells with two recombinant viruses, each synthesizing two BTV proteins (see Chapter 2 for further details on multiple expression vectors). As already mentioned, expression of the entire coding region of poliovirus type 3 gives rise to the assembly of non-infectious, intact virions containing VP0, VP1 and VP3 (Urakawa *et al.*, 1989). When the poliovirus capsid coding region alone was expressed, the only protein synthesized was an uncleaved precursor, thus demonstrating the requirement for the poliovirus encoded protease to complete the proteolytic processing events (Urakawa et al., 1989).

Although most of the studies on quaternary structure formation have focused on virus proteins, there are also examples of other recombinant proteins that have formed complex structures in insect cells. These include: dimeric assembly of the human macrophage-colony-stimulating factor (Maiorella *et al.*, 1988) and platelet-derived growth factor (Giese *et al.*, 1989); heterodimer (α and β chain) formation of the extracellular domain of the insulin receptor (Sissom and Ellis, 1989); dimeric assembly of the extracellular domain of the nerve growth factor receptor (Vissavajjhala and Ross, 1990) and the transferrin receptor (Domingo and Trowbridge, 1988); and tetrameric assembly of rat tyrosine hydroxylase (Forstová *et al.*, 1989).

3.7 EXPRESSION OF VIRAL GENES

The baculovirus system has been used extensively to express virus structural and non-structural genes. Some of these have been used as examples in the sections covered above, for example, virus glycoproteins (section 3.2). Many of the virus structural proteins have been synthesized to very high levels and virtually all published examples have proved to be antigenic, giving rise to high-titre antisera. In many cases challenge experiments with live virus have shown that the insect cell-derived proteins are capable of inducing protective, neutralizing antibodies. Examples of virus structural and non-structural proteins synthesized in insect cells are given in Table 3.3, except glycoproteins which have been included in Table 3.1.

3.8 EXPRESSION OF BACTERIAL AND FUNGAL GENES

The sections above refer almost exclusively to the expression of eukaryotic or virus genes. A number of bacterial and fungal genes have also been expressed very successfully in insect cells. Examples include: *Escherichia coli* β-galactosidase (Pennock *et al.*, 1984; and subsequently by many workers), β-glucuronidase (Luckow and Summers, 1989) and chloramphenicol acetyltransferase (Luckow and Summers, 1988b, 1989); *Bacillus thuringiensis* (subspecies *Kurstaki* HD73) delta-endotoxin (Martens *et al.*, 1990; Merryweather *et al.*, 1990); and *B. anthracis* protective antigen (Iacono-Conners *et al.*,

Table 3.3 Examples of virus structural and non-structural proteins synthesized in insect cells[1]

Protein	*Virus*	*Structural (S), Non-structural (NS)*	*Reference*
DNA polymerase	Adenovirus type 2	NS	Watson and Hay (1990)
VP7 (core)	Bluetongue virus (type 10)	S	Oldfield *et al.* (1990)
NS3 and NS3A	Bluetongue virus (type 10)	NS	French *et al.* (1989)
Four proteins	Bluetongue virus (type 10)	S[2]	French *et al.* (1990)
Two core proteins	Bluetongue virus (type 10)	S[2]	French and Roy (1990)
NS1 gene	Bluetongue virus (type 10)	NS[2]	Urakawa and Roy (1988)
E2 ORF	Bovine papillomavirus	NS	Tada *et al.* (1988)
I gene	Cauliflower mosaic virus	NS	Vlak *et al.* (1990)
Core	Dengue virus	S	Makino *et al.* (1989)
gag proteins	Feline immunodeficiency virus	S[2]	Morikawa *et al.* (1991)
Polyprotein	Foot-and-mouth-disease-virus	NS, S	Roosien *et al.* (1990)
Capsid proteins	Hepatitis A virus	S	Harmon *et al.* (1988)
Core antigen	Hepatitis B virus	S[2]	Takehara *et al.* (1988) Hilditch *et al.* (1990)
DNA polymerase	Herpes simplex virus	NS	Marcy *et al.* (1990)
Helicase–primase	Herpes simplex virus (type 1)	NS	Dodson *et al.* (1989)
Protease	HIV	NS	Overton *et al.* (1989)
p17gag, p24gag	HIV	S	Overton *et al.* (1989)
gag	HIV	S	Madisen *et al.* (1987)
Nucleoproteins	Influenza A and B viruses	S	Rota *et al.* (1990)
Three polymerase proteins	Influenza virus	NS	St Angelo *et al.* (1987)
VP1 and VP2	Parvovirus (human)	S[2]	Brown *et al.* (1991)
Polyprotein	Poliovirus (type 3)	NS, S[2]	Urakawa *et al.* (1989)
Large T antigen	Polyoma virus	NS	Rice *et al.* (1987)
Nucleoprotein	Rabies virus	S	Préhaud *et al.* (1990)
VP4 (outer capsid)	Rotavirus	S	Nishikawa *et al.* (1989)
Major capsid (VP6)	Rotavirus	S[2]	Estes *et al.* (1987)
Polyprotein	Sindbis virus	NS, S	Oker-Blom and Summers (1989)
Nucleoprotein	Snowshoe hare bunyavirus	S	Urakawa *et al.* (1988)
NSs	Snowshoe hare bunyavirus	NS	Urakawa *et al.* (1988)
Large T antigen	SV40	NS	Lanford (1988) O'Reilly and Miller (1988)
Small t antigen	SV40	NS	Jeang *et al.* (1987b)

[1] Glycoproteins are included in Table 3.1.
[2] Tertiary or quaternary structures formed.

1990). Finally, the fungal qa-1F activator protein from *Neurospora crassa* has been successfully synthesized in insect cells, where other systems (bacterial, yeast) have failed due to the cytotoxic nature of this protein (Baum *et al.*, 1987).

3.9 POST-TRANSCRIPTIONAL PROCESSING

There is very little published information concerning the ability of the baculovirus system to cope with the expression of intron-containing genes. Other processing events such as polyadenylation and 5′ capping of mRNAs are known to occur. One report has tentatively concluded that the insect cell can distinguish splice acceptor–donor sites. In this study, a segment of SV40 genomic DNA encoding the large T and small t antigens was expressed in a baculovirus vector (Jeang *et al.*, 1987b). Only small t antigen was detected in the insect cells and the authors concluded that this demonstrated that the splice site for small t antigen was preferentially utilized. To date, there have been no published reports that an intron-containing gene can be accurately spliced to give the correct mRNA for faithful translation.

Construction of transfer vectors containing the foreign gene 4

4.1 INTRODUCTION

This chapter describes all of the techniques required for inserting a foreign coding sequence into a baculovirus transfer vector and preparing DNA which is suitable for performing co-transfection experiments to derive recombinant virus. Most of the techniques presented are not unique to the baculovirus expression system, and the reader who is experienced with molecular biology may not necessarily agree with all of our protocols. The methods in this chapter, however, represent the everyday procedures used in our own laboratories and are presented in the knowledge that they do work. Alternative and additional procedures may be gleaned from the excellent laboratory manuals edited by Berger and Kimmel (1987) and Sambrook *et al.* (1989).

4.2 ISOLATION OF FOREIGN GENE CODING SEQUENCES

4.2.1 SOME GENERAL GUIDELINES

The starting point is a plasmid/phage vector with the foreign gene coding sequences inserted at a convenient restriction enzyme (RE) site to enable release and subsequent purification. If the insertion of the DNA into the vector has resulted in the destruction of the RE recognition sequence then it will be necessary to use other RE sites which flank your foreign gene coding sequence. This may result in the addition of plasmid sequences to the foreign gene. In consequence, the amount of 5′ and 3′ non-coding sequences associated with the coding region may be variable, and also dependent on the source of the foreign gene (genomic DNA, cDNA, PCR product, etc.). The effect of these extra sequences on expression in the baculovirus system is uncertain.

It is generally accepted that 5′ non-coding sequences rich in GC nucleotides are undesirable, although there are few experimental data to confirm this assumption. It is essential, however, to ensure that there are not additional translation initiation codons (AUG) before the authentic codon at the start of your foreign gene coding sequences. These extra AUG codons must be removed using exonuclease deletion, site-directed mutagenesis or PCR methods. Similar techniques can also be used to add an AUG start codon if your gene sequences lack one, or alternatively you can use one of the fusion protein vectors (described in Chapter 2) that use the polyhedrin AUG.

The requirement for a Kozak consensus sequence (Kozak, 1986) is also debatable. In relation to the AUG start codon, the normal AcMNPV polyhedrin gene has an A at position −3; however, foreign gene coding sequences lacking this arrangement have been successfully expressed to high levels, while others with an ideal Kozak consensus have been poorly expressed. With the advent of PCR techniques it is now feasible to tailor the foreign gene coding sequences quickly. If this method is available, and the foreign gene is not too large, then a copy may be produced which precisely mimics the translation initiation region of the polyhedrin gene (see Innis *et al.* (1990) for more information). However, we advise that before spending a great deal of time and effort altering the 5′ sequences, you first of all test expression using the existing construction.

If you desire your foreign protein to be secreted or to be membrane targeted, you must ensure that the appropriate signal peptide and anchor regions are present (see Chapter 3 for more details). The length and nature of the 3′ non-coding sequences have less importance than those at the 5′ end and may be discounted, as long as you have a 3′ translational stop codon at the end of your foreign gene coding sequences.

The other problem which must be considered is the compatibility of the DNA ends produced by the REs required to remove the foreign gene coding sequence from the host vector, with the cloning sites within the baculovirus transfer vector. Most baculovirus transfer vectors utilize *Bam*HI or *Bgl*II insertion sites; some others have polylinker cloning regions (see Chapter 2). In our laboratories we have been inclined to avoid the use of transfer vectors containing polylinker cloning sites; this is a prejudice dating from the period when the importance of the 5′ non-coding region was uncertain. The availability of transfer vectors with multiple sites for inserting the coding sequences greatly simplifies the cloning procedure. When these vectors are not available, or unsuitable because of identical RE sites in the coding sequences, the ends of the foreign DNA must be modified by the addition of synthetic linkers.

4.2.2 ISOLATION OF DNA FRAGMENTS FROM AGAROSE GELS

Having selected the baculovirus transfer vector (see Chapter 2) and checked that the foreign gene coding sequence is suitable to be inserted into the vector, the next step is to isolate and purify the DNA fragment encoding the foreign gene. Several methods are available for the purification of DNA fragments from agarose gels and these have been summarized by Sambrook *et al.* (1989). We find that the most convenient of these is the use of phenol to remove agarose from the DNA.

Required:

10× restriction enzyme buffer
Restriction enzyme(s)
Low-gelling temperature agarose (e.g. SeaPlaque by FMC Bio-Products or Sigma type VII)
10× TBE (108 g Tris, 55 g boric acid, 7.8 g EDTA per litre)
Phenol equilibrated in water (phenol/H20)
Phenol/chloroform (50:50) equilibrated in 50 mM Tris-HCl, pH 8.0
3 M sodium acetate
Ethanol (100% and 75%)
TE buffer (10 mM Tris-HCl, pH 8.0, 1 mM EDTA)
1 M Tris-HCl, pH 8.0
10 mg/ml ethidium bromide (take appropriate safety precautions)

Method:

1. Digest the vector containing the foreign gene coding sequences with suitable REs which flank the insert. In general, 10 μg of DNA is sufficient, but this will depend on the efficiency of extracting DNA fragments from agarose gels. The recovery is also influenced by the size of the DNA fragment; small fragments (< 500 nucleotides) are usually more difficult to purify.

Plasmid DNA (ca. 10 μg)	× μl
10× RE buffer	10 μl
RE(s) (10 units/μl)	2 μl
Sterile water	to 100 μl

Incubate at 37°C for at least 2 h and then monitor the extent of digestion by analysing 100–200 ng DNA in an agarose mini-gel.

If necessary, add more enzyme and continue the incubation until the reaction is complete.
* Choose enzymes that give a clean separation of fragments in an agarose gel. If necessary select a second or third enzyme which cleaves the plasmid backbone and results in better DNA fragment resolution.

2. Prepare a 0.6–1% (w/v) low-gelling temperature agarose gel in TBE buffer with a single large well. Equilibrate the gel and running buffer at 4°C to ensure the complete setting of the agarose. Fractionate the digested DNA and stain the gel with 0.5 μg/ml ethidium bromide (this may be included in the running buffer). Excise the desired DNA fragment using a clean surgical blade, trim excess agarose and place in a 1.5 ml microfuge tube.
3. Melt the agarose at 65°C and dilute to 0.3% with TE. Add Tris-HCl, pH 8.0 to 100 mM. If the agarose solidifies, re-melt at 65°C and then equilibrate at 37°C.
4. Extract with an equal volume of phenol/H_2O for 30 min at 37°C.
* This is conveniently performed using a shaking incubator normally reserved for growing bacterial cultures.
5. Separate the phases at high speed in a microfuge for 5 min and briefly re-extract the aqueous phase with phenol/chloroform. Separate the phases as before.
6. Add sodium acetate to a final concentration of 0.3 M, followed by 2 volumes of 100% ethanol. Precipitate the DNA at −20°C for 1 h.
7. Pellet the DNA in a microfuge for 30 min (preferably at 4°C); this extended centrifugation step ensures complete recovery of the DNA fragment, particularly if it is small (< 500 nucleotides).
8. Wash the pellet twice with 75% ethanol, remove residual ethanol with a drawn-out Pasteur pipette and briefly air dry the DNA.
9. Resuspend the DNA in 20 μl TE buffer and assess the concentration by analysing 2 μl in a mini agarose gel, with appropriate DNA size markers (e.g. lambda DNA digested with *Hin*dIII).
* The above method produces clean DNA which may be ligated with the transfer vector (see below). An alternative method, although more costly, is the use of Geneclean (Bio 101, California, USA). This method does not require organic extractions and precipitations and can be accomplished in about 30 min (after excising the DNA fragment from the gel).

4.3 MODIFYING THE ENDS OF DNA MOLECULES

If convenient RE sites flanking the foreign gene coding sequences are not available, it may be necessary to convert the DNA termini to blunt ends using one of various methods available. The two most common methods involve Mung Bean nuclease or *E. coli* large fragment of DNA polymerase (Klenow) enzyme. Neither approach is entirely satisfactory, since it is impossible to treat efficiently all of the available ends.

4.3.1 MUNG BEAN NUCLEASE

Required:

5× Mung bean buffer (150 mM sodium acetate, pH 5.0, 250 mM NaCl, 5 mM $ZnCl_2$, 25% glycerol)
1× Mung bean dilution buffer (10 mM sodium acetate, pH 5.0, 0.1 mM zinc acetate, 1 mM cysteine, 0.001% Triton X-100, 50% glycerol)
Mung bean nuclease (e.g. from Gibco-BRL)
Phenol/chloroform (50:50) in 50 mM Tris-HCl, pH 8.0
3 M sodium acetate
Ethanol (100% and 75%)

Method:

1. Resuspend the DNA (3 μg) to be treated in 200 μl water and add 50 μl 5× mung bean buffer.
2. Add 9 units mung bean nuclease (previously diluted with 1× dilution buffer) and incubate for 30 min at 30°C. Stop the reaction with 250 μl phenol/chloroform.
3. Mix thoroughly and separate the phases using a high-speed microfuge. Remove the aqueous phase to a fresh tube.
4. Add 25 μl 3 M sodium acetate and 500 μl 100% ethanol. Mix and then pellet the DNA for 10 min at high speed in a microfuge. wash the pellet twice with 75% ethanol.
5. Dry the pellet and resuspend the DNA in 30 μl water.

4.3.2 KLENOW FILL-IN

Required:

RE with appropriate 10× buffer
2 mM dNTP stock (all four), stored at −20°C
10% w/v Sodium dodecyl sulphate (SDS)
0.5 M EDTA

Phenol/chloroform (50:50) in 50 mM Tris-HCl, pH 8.0
3 M sodium acetate
Ethanol (100% and 75%)

Method:

1. Digest the DNA (5 μg) with the required RE.
2. Prepare the following reaction:
   digested DNA (5 μg) 50 μl
   2 mM dNTPs (all four) 5 μl
   Klenow (1–5 units) × μl
   Incubate at room temperature for 10 min.
3. Stop the reaction by adding SDS to 1% and EDTA to 10 mM.
4. Extract with phenol/choroform twice and ethanol precipitate the DNA. Wash the DNA pellet in 75% ethanol.
5. Resuspend the DNA in 50 μl TE buffer and purify the appropriate DNA fragment containing the foreign gene as described in section 4.2.2.

4.4 PREPARATION OF THE TRANSFER VECTOR

If you have any doubts about which baculovirus transfer vector to use, consult Chapter 2 before proceeding further. We have included, at the end of this chapter, detailed RE maps and listings of the RE sites in pAcYM1 (single expression vector) and pAcUW3 (dual expression vector). This should be of considerable assistance to users of the system who wish to plan appropriate RE digestions of their recombinant plasmids to confirm, for example, that the foreign coding sequence has been inserted in the correct orientation relative to the baculovirus gene promoter. In this section we assume that a supply of the purified plasmid DNA is available. If the transfer vector has to be amplified prior to use, because it was provided as a bacterial culture, then consult section 4.9 for details of plasmid purification techniques.

Required:

Reagents described in 4.2 for digestion of DNA with RE and subsequent phenol extraction and ethanol precipitation
250 mM EDTA
10% w/v SDS
Calf intestinal phosphatase (CIP) (molecular biology grade; e.g. Boehringer Mannheim)

Method:

1. Digest 5–10 μg of the required transfer vector with the appropriate RE, as described in section 4.3.2

2. When the digestion is complete add 5–10 units CIP directly to the reaction mixture and continue the incubation for 15 min.
   * Although most suppliers usually recommend precipitating the digested plasmid at this point and redissolving in a buffer optimal for CIP activity, in our experience the enzyme functions perfectly well in most RE buffers. Our one concession is to dilute high salt restriction enzyme buffers (e.g. 100+ mM NaCl) to medium salt conditions (50 mM).
3. Add EDTA to 25 mM and SDS to 1% and extract with an equal volume of phenol/chloroform for 1 min. Separate the phases using highspeed in a microfuge and briefly re-extract the aqueous phase with phenol/chloroform.
4. Add sodium acetate to 0.3 M and two volumes of ethanol. It is not usually necessary to chill the DNA prior to centrifugation.
5. Pellet the DNA for 10 min in the microfuge and wash the pellet twice with 75 % ethanol. After drying, resuspend the DNA in 50–100 μl TE buffer and assess the concentration using an agarose mini-gel. Adjust to approximately 100 ng/μl with TE buffer. Store at −20°C.

4.5 DNA LIGATIONS

After purifying the relevant DNA fragment containing the foreign gene coding sequences for expression, modifying the ends where necessary, and preparing the appropriate baculovirus transfer vector, the next step is to join them together using DNA ligation techniques. Unlike plasmid vectors such as pUC18 and 19, which have a blue/white selection system for identifying recombinant plasmids containing inserted DNA sequences, or pBR322, which has ampicillin and tetracycline resistance genes to aid recombinant selection, the baculovirus transfer vectors lack such facilities. In consequence, the selection of transfer vectors containing foreign gene coding sequences may present a problem. It is essential that the digested, CIP-treated transfer vector gives a very low background level of colonies after transformation of bacteria, to ease the identification of recombinant plasmids. The inclusion of appropriate controls in the DNA ligations will permit the estimation of the level of non-recombinant contamination in the final colony count.

Required:

10× ligation buffer (0.66 M Tris-HCl, pH 7.5, 50 mM $MgCl_2$, 50 mM Dithiothreitol (DTT)), stored at −20°C
10 mM ATP in water (stored in aliquots at −20°C)
T4 DNA ligase (5 units/μl)

Sterile water
RE-digested, CIP-treated baculovirus transfer vector (see section 4.4)
Purified DNA fragment containing foreign gene coding sequences, for insertion into transfer vector (see section 4.2)

Method:

1. Set up the ligation reaction:

RE-digested, CIP-treated transfer vector (50 ng)	0.5 μl
Purified DNA insert (1–10× molar excess over vector)	× μl
10× ligation buffer	2 μl
10 mM ATP	2 μl
T4 DNA ligase (1 unit)	0.2 μl
Sterile, water	to 20 μl

   * The volume of the ligation reaction may have to be increased if the concentration of the DNA insert is particularly low. We work on the basis that the volume of the insert should not be greater than 50% of the total ligation reaction volume.
   * It is worthwhile setting up two or three separate ligation reactions, each with differing amounts of insert DNA with respect to transfer vector DNA.
2. Set up control ligation reactions: (1) without the insert or ligase; (2) without the insert, but with ligase. These serve as a way of quickly determining whether the ligation of the insert to the vector has been satisfactory (after transformation of bacteria).
3. For sticky-end ligations, incubate at 16°C for 12–16 h. Alternatively, incubate at room temperature (21°C) for 3–4 h. Blunt-end ligations seem to be accomplished more efficiently at room temperature (12–16 h).
4. When the ligation reactions are complete they should be placed on ice if they are to be used to transform bacteria immediately (or within a few hours); should competent bacterial cells not be available the reactions may be stored at −20°C.

4.6 TRANSFORMATION OF BACTERIA

The preparation of competent bacterial cells for transformation with plasmids is a routine procedure and will not be described in this manual. Appropriate methods may be found in Sambrook *et al.* (1989). It is worth preparing frozen stocks of competent cells for subsequent use. Most of the commonly used *E. coli* strains may be used, e.g. DH1, JM101, JM105, HB101. If you experience any problems inserting your foreign gene into the transfer vector, you could try SURE™ cells (Trademark of Stratagene Ltd) which help prevent rearrangements and deletions occurring. We recommend the following rapid method for transforming these cells with baculovirus transfer vectors.

Required:

A frozen stock of competent *E. coli* cells (JM105 or other suitable strain, Sambrook *et al.*, 1989)

Luria broth (LB; 0.5% NaCl, 1.0% bactotryptone, 0.5% yeast extract, 2 mM NaOH, autoclaved before use)

LB-agar (1.5% bactoagar in LB) plates containing ampicillin (50 μg/ml) (LB-agar^{amp50})

42°C water bath

37°C incubator

95% alcohol and a glass spreader for bacterial cells

Method:

1. Thaw the appropriate quantity of frozen, competent bacterial cells on ice for about 1 h; mix thoroughly prior to use. You will require 100–200 μl per ligation reaction.
2. Chill the ligation reactions (see section 4.5), add 100 μl competent cells and leave on ice for 10 min.
3. Heat shock the cells at 42°C for 2 min, then return to room temperature.
4. Add 1 ml LB and incubate at 37°C for 1 h, with gentle agitation.
5. Spread 100 μl of the cells onto a LB-agar^{amp50} plate.
6. Pellet the remaining cells at low speed in a microfuge and resuspend in approximately 100–150 μl LB. Spread this cell concentrate on another LB-agar^{amp50} plate.
   * The plating of two concentrations of bacterial cells provides a 'safety net' for inefficient ligation reactions and usually ensures that sufficient colonies are obtained for subsequent screening.
7. Invert and incubate all plates at 37°C overnight.
8. Count the numbers of colonies on all plates and compare results for those ligations with/without DNA inserts. This provides a quick check on the likely success of the experiment. A ratio of greater than 5:1 (with:without insert) usually means that the transfer vector contains the required sequences.
9. The colonies from the plate with inserted DNA may be screened by preparing mini-prep DNA (section 4.7.2) for RE analysis (section 4.8).

4.7 SCREENING FOR RECOMBINANT BACULOVIRUS TRANSFER VECTORS

If the number of colonies obtained after transformation is favourable, then the quickest way to screen for the desired insert within the transfer vector is to isolate plasmid DNA using the mini-prep method (section 4.7.2). The DNA can then be characterized using RE and agarose gel electrophoresis. On occasion, however, a particular sequence may prove difficult to insert into the transfer vector and thus yield a low number of recombinant plasmids. This can sometimes happen if the gene to be inserted is large and you are using one of the larger transfer vectors (see Chapter 2 for more details). In these cases it may be advantageous to perform colony hybridization analysis (section 4.7.1) (Grunstein and Hogness, 1975) to identify those colonies harbouring the foreign gene coding sequences.

4.7.1 COLONY HYBRIDIZATION

The method described here involves transferring colonies to a hybridization membrane and a replica library plate. It is possible to lift the colonies directly onto a filter (thus avoiding replica plating) and then process them for hybridization; this method uses more hybridization membranes and is only feasible if the colonies on each dish are well separated.

Required:
- Hybridization membranes (circular, 9 cm diameter with printed grid pattern; nitrocellulose or nylon)
- Sterile toothpicks
- LB-agar^{amp50} plates (see section 4.6)
- 4 × 15 cm diameter Petri dishes, each containing several layers of Whatman 3 MM paper soaked with*:
  - Dish 1: 0.2 M NaOH
  - Dishes 2 and 3: 0.5 M Tris-HCl, pH 7.4
  - Dish 4: 0.5 M Tris-HCl, pH 7.4, 1.5 M NaCl
  - * Do not have excessive buffer in the dishes.
  - Blunt forceps
  - 80°C and 65°C incubators
  - 9 cm Petri dishes
  - Whatman 3 MM paper
  - Colony hybridization solution (6× SSC, 0.2% BSA, 0.2% Ficoll, 0.2% polyvinylpyrollidone)
  - 3× SSC, 0.1% SDS
  - (20× SSC is 3 M NaCl, 0.3 M tri-sodium citrate)

[^{32}P]-labelled probe specific for the foreign gene coding sequence (see Sambrook *et al.*, 1989)
Facilities for autoradiography

Method:

1. Place a hybridization membrane onto an LB-agar^{amp50} plate and ensure that all air-bubbles are removed.
2. Streak each putative recombinant colony onto the membrane with a sterile toothpick and then replica plate onto a normal LB-agar^{amp50} plate (which has an identical grided pattern drawn on its base with a marker pen); the second plate provides a library for later recovery of the recombinant colonies.
   * It is wise to include negative (bacteria transformed with the unmodified transfer vector) and positive (bacteria transformed with a plasmid containing the foreign coding sequences) controls.
3. Grow the colonies for about 5 h at 37°C, until a film of bacteria on the membrane is clearly visible. The library plate should be incubated at 37°C for 12–16 h, sealed with parafilm and stored at 4°C.
   * We have found that if the colonies are grown on the membrane for only a limited period then the subsequent lysis procedure is accomplished efficiently and, more importantly, the background hybridization signal in non-recombinant colonies is very low.
4. Place the hybridization membrane, colony side uppermost, onto Dish 1 for 8 min to lyse the cells.
   * It is important that the liquid does not come into contact with the upper surface or the cells will wash into one another.
   * Lysis of the bacterial cells is accompanied by the colonies adopting a shiny appearance.
5. Transfer the membrane to Dishes 2 and 3 for 2 min in succession.
6. Transfer the membrane to Dish 4 for 4 min.
7. Dry the membrane on two layers of Whatman 3MM paper at 37°C for 10 min, then bake at 80°C for 2 h. The membrane may be stored indefinitely after this step has been completed.
8. Transfer the membrane to a plastic Petri dish (9 cm diameter) and add 25 ml colony hybridization solution (three membranes may be treated per dish). Incubate at 65°C for 3–4 h.
9. Mix 15 ml fresh colony hybridization solution with the boiled, denatured radioactive probe. Replace the prehybridization solution with this mixture and incubate at 65°C overnight.
10. Remove the solution containing the radioactive probe and wash the filter four times (30 min) with 3× SSC, 0.1% SDS at 65°C.

11. Seal the filter in a plastic bag and expose to X-ray film using standard autoradiography techniques.
    * If the membrane is kept moist it is possible to rewash it if the background hybridization signal is too high, preventing easy identification of positive colonies.
12. Develop the exposed X-ray film and use it to identify recombinant transfer vectors containing the foreign coding sequences destined for expression in the baculovirus.
13. Identify the recombinant transfer vector on the stored library plate.
14. Proceed to section 4.7.2.

4.7.2 RAPID ISOLATION OF BACTERIAL PLASMID DNA (MINI-PREPS)

This method is based on that of Maniatis *et al.* (1982).

Required:

LB containing ampicillin (50 μg/ml) (LB^{amp50}, see section 4.6)
Glucose solution (50 mM glucose, 25 mM Tris-HCl, pH 8.0, 10 mM EDTA)
Alkaline lysis buffer (0.2 M NaOH, 1% w/v SDS, freshly prepared)
Acetate solution (add 11.5 ml glacial acetic acid and 28.5 ml water to 60 ml 5 M potassium acetate)
Ribonuclease A (10 mg/ml in water, heated to 100°C for 10 min and stored at −20°C)
Phenol/chloroform (50:50, equilibrated in 50 mM Tris-HCl, pH 8.0)
TE buffer (10 mM Tris-HCl, pH 8.0, 1 mM EDTA)

Method:

1. Inoculate 2 ml volumes of LB^{amp50} with the required number of bacterial colonies using a sterile wooden toothpick or bacterial loop. Grow the bacteria with vigorous shaking for 14–18 h at 37°C.
   * It is possible to grow the cultures for only 6–7 h, and still produce enough plasmid DNA for subsequent analysis. Most baculovirus transfer vectors give very high yields of DNA.
2. Pour 1.5 ml of the culture into a microcentrifuge tube. Store the remainder of the culture at 4°C for later use in maxi-preps (section 4.9). If the culture is not to be used for some time (weeks) then store in 15% glycerol at −70°C.
3. Pellet the cells at low speed in a microfuge and carefully remove, and discard safely, the supernatant fraction from each tube.
4. Resuspend the cells in 100 μl glucose solution.

5. Add 200 μl alkaline lysis buffer and 1 μl 10 mg/ml ribonuclease A, mix gently and allow to stand on ice for 5 min.
6. Addd 150 μl acetate solution, mix gently and stand on ice for 10 min.
7. Pellet the chromosomal DNA and other debris for 10 min at high speed in a microfuge and then remove the supernatant fraction (containing plasmid DNA) into a fresh microfuge tube.
8. Add 300–400 μl phenol/chloroform to the plasmid DNA and mix vigorously (vortex) for 1 min. Separate the phases by spinning in a microfuge for 5 min and transfer the upper, aqueous phase to a fresh microfuge tube.
9. Add two volumes of ethanol and chill at −20°C for 30 min. Recover the DNA by centrifugation for 10 min in a microfuge at high speed.
10. Remove the ethanol and wash the DNA pellet twice with 75% ethanol. Remove all traces of the ethanol using a drawn-out Pasteur pipette and briefy air-dry the pellet of DNA. Resuspend the DNA in 20 μl TE buffer and store at −20°C until required.
    * The above method produces reasonably clean DNA which is suitable for analysis with REs. We have not found any problems with contaminating nucleases. There have been some suggestions that the DNA is also suitable for performing co-transfections to derive recombinant baculoviruses (see Chapter 7).

4.8 ANALYSIS OF RECOMBINANT TRANSFER VECTORS

Using the above method, several micrograms of plasmid DNA may be obtained from 1.5 ml of bacterial culture. This is more than adequate for most RE analyses, to determine which clones have the foreign gene inserted in the correct orientation with respect to the polyhedrin (p10, basic protein, etc.) promoter. The combination of RE to be used for digesting the plasmid DNA will depend on the availability of suitable sites within the inserted foreign gene. To assist in this choice we have provided a detailed map and complete list of the RE which recognize six nucleotides for the plasmid transfer vectors pAcYM1 (Matsuura *et al.*, 1987) (Figure 4.1 and Table 4.1) and pAcUW3 (Weyer and Possee, 1991) (Figure 4.2 and Table 4.2). The complete nucleotide sequence of the AcMNPV *Eco*RI I fragment, from which all polyhedrin-based transfer vectors have been derived, has been published (Possee *et al.*, 1991) and is available from Genebank (Accession number M75679).

When planning your RE digestions, consider RE combinations that both identify the presence and orientation of the insert and the copy number (it is not uncommon for two fragments to ligate together

Table 4.1 Restriction enzyme sites in pAcYM1 that recognize a 6-nucleotide sequence[1]

Name	*Position*								
*Aat*II	(3)	1500	4446	7068					
*Acc*I	(5)	1443	2948	3233	5533	6518			
*Afl*III	(17)	67	263	1452	1607	1609	1757	1939	2469
		2681	3388	4516	4666	4771	5540	5665	6455
		8875							
*Aha*II	(12)	1497	1730	1864	1917	3484	4443	4643	4996
		5263	6762	7065	7447				
*Apa*I	(1)	1399							
*Apa*LI	(7)	424	2454	2465	6253	6819	7315	8561	
*Asu*II	(3)	521	2057	6188					
*Ava*I	(3)	163	1901	2266					
*Ava*II	(5)	203	1516	2634	7623	7845			
*Bal*I	(3)	335	1506	5755					
*Bam*HI	(1)	4095							
*Ban*I	(9)	1729	4968	4995	5199	5250	5262	6761	8034
		9131							
*Ban*II	(1)	1399							
*Bbv*I	(25)	1141	1335	1881	2290	2960	3631	3795	3829
		4399	5097	5259	5314	5759	6682	6755	6943
		7553	7944	8247	8453	8456	8546	8965	8983
		9064							
*Bcl*I	(1)	2232							
*Bgl*I	(2)	6752	7869						
*Bin*I	(17)	35	883	1220	2734	3752	4090	4103	5456
		5465	7348	7352	7559	8132	8133	8229	8231
		8317							
*Bsm*I	(3)	595	2673	2926					
*Bsp*1286	(14)	428	1146	1236	1338	1399	2458	2469	4753
		5253	6257	6823	7319	7404	8565		
*Bsp*MI	(3)	936	1032	1537					
*Bsp*MII	(2)	2822	6286						
*Bst*EII	(1)	923							
*Bst*XI	(3)	749	908	3156					
*Bst*NI	(8)	332	487	977	6644	8716	8729	8850	9138
*Cla*I	(3)	251	5137	5460					
*Dra*I	(13)	147	1484	1573	1835	2886	3418	4255	5580
		5589	6451	7409	8101	8120			
*Dra*II	(3)	1395	1396	7007					
*Eae*I	(10)	111	333	801	1504	1932	5753	6362	6608
		7594	9036						
*Eco*K	(2)	5608	7316						
*Eco*RI	(2)	5608	7316						
*Eco*RV	(1)	4000							
*Fok*I	(9)	865	1421	2168	2594	6662	6906	7548	7835
		8016							

Name	*Position*
*Fsp*I	(5) 2085 3648 5685 6742 7764
*Hae*II	(11) 1733 4125 4999 5164 5266 5308 5844 6124 6765 8635 9005
*Hga*I	(19) 824 922 1022 1388 1872 1925 1991 2455 2693 3135 3492 3955 4225 4651 6408 6898 7455 8185 8763
*Hgia*I	(11) 428 1146 1236 1338 2458 2469 6257 6823 7319 7404 8565
*Hinc*II	(8) 1888 1915 2843 2949 3234 5534 6340 6470
*Hind*III	(2) 4528 5574
*Hpa*I	(3) 2843 6340 6470
*Hph*I	(15) 935 1160 1163 2598 3172 3868 4929 5935 6968 6977 7260 7295 7501 7917 8144
*Mbo*II	(20) 202 205 632 1243 1552 3057 3487 3490 3493 3685 3781 4629 6399 6718 7205 7314 7392 8147 8218 9009
*Mlu*I	(2) 2469 3388
*Nae*I	(1) 3772
*Nar*I	(4) 1730 4996 5263 6762
*Ncl*I	(6) 407 6916 6951 7451 7802 8498
*Nde*I	(1) 6814
*Nsl*I	(2) 2673 3173
*Nsp*BII	(15) 870 1309 1360 1966 2639 3644 5088 5268 6111 6692 6886 7351 8292 8537 9055
*Pvu*I	(2) 6723 7618
*Pvu*II	(4) 1309 6111 6692 9055
*Sac*II	(1) 871
*Sal*I	(3) 2947 3232 5532
*Sca*I	(1) 7506
*Sfa*NI	(15) 1076 2381 2661 3161 4929 5676 5690 6777 6779 6853 6909 7305 7516 7745 8797
*Sna*BI	(1) 4165
*Sph*I	(1) 2135
*Ssp*I	(11) 274 587 2046 2996 3414 4259 4456 5329 5585 6099 7182
*Sty*I	(1) 1501
*Xho*I	(1) 1901
*Xho*II	(7) 4095 7340 7357 8125 8137 8223 8234
*Xma*III	(1) 6362
*Xmn*I	(4) 1247 4706 6290 7387

[1] Matsuura *et al.* (1987); Possee *et al.* (1991); see also Figure 4.1.

prior to insertion into the transfer vector). Ideally, you need to select a RE that digests the transfer vector once (*Eco*RV or *Xho*I in pAcYM1 or pAcRP23 are good choices) and one that digests your inserted gene once in an asymmetrical manner, so that orientation of the gene with respect to the polyhedrin promoter can be ascertained.

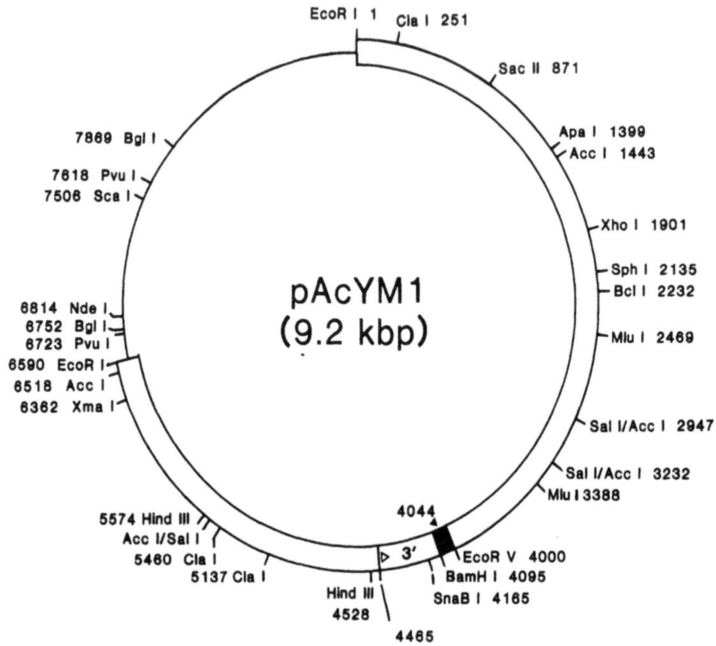


```
        3970      3980      3990      4000      4010      4020
TCACAAACTGGAAATGTCTATCAATATATAGTTGCTGATATCATGGAGATAATTAAAATG
                                    EcoR V
        4030      4040   T  4050      4060      4070      4080
ATAACCATCTCGCAAATAAATAAGTATTTTACTGTTTTCGTAACAGTTTTGTAATAAAAA

        4090      4100      4110      4120      4130      4140
AACCTATAAATACGGATCCGGTTATTAGTACATTTATTAAGCGCTAGATTCTGTGCGTTG
             BamH I
        4150      4160      4170      4180      4190      4200
TTGATTTACAGACAATTGTTGTACGTATTTTAATAATTCATTAAATTTATAATCTTTAGG
                     SnaB I
```

Figure 4.1 Restriction enzyme map of pAcYM1 (Matsuura *et al.*, 1987). See Table 4.1 for a detailed listing of all REs that recognize a 6-nucleotide sequence. The plasmid backbone (pUC derivative) is indicated by a single line and the AcMPNV sequences by a double line. The *Bam*HI insertion site for foreign gene sequences is at position 4095. The polyhedrin promoter is indicated by a solid box. The site of transcription initiation is shown at position 4044 (solid triangle) and the 3′ end of the mRNA is shown at position 4465 (open triangle). The sequence spanning the *Bam*HI insertion site is shown beneath the circular map.

Table 4.2 Restriction enzyme sites in pAcUW3 (dual-expression vector) that recognize a 6-nucleotide sequence[1]

Name	*Position*								
*Aat*II	(3)	1500	5166	7788					
*Acc*I	(5)	1443	2948	3233	6253	7238			
*Afl*III	(18)	67	263	1452	1607	1609	1757	1939	2469
		2681	3388	4663	5236	5386	5491	6260	6385
		7175	9595						
*Aha*II	(12)	1497	1730	1864	1917	3484	5163	5363	5716
		5983	7482	7785	8167				
*Apa*I	(1)	1399							
*Apa*LI	(7)	424	2454	2465	6973	7539	8035	9281	
*Asu*II	(3)	521	2057	6908					
*Ava*I	(3)	163	1901	2266					
*Ava*II	(6)	203	1516	2634	4569	8343	8565		
*Bal*I	(3)	335	1506	6475					
*Bam*HI	(1)	4815							
*Ban*I	(9)	1729	5688	5715	5919	5970	5982	7481	8754
		9851							
*Ban*II	(1)	1399							
*Bbv*I	(26)	1141	1335	1881	2290	2960	3631	3795	3829
		4105	5119	5817	5979	6034	6479	7402	7475
		7663	8273	8664	8967	9173	9176	9266	9685
		9703	9784						
*Bcl*I	(2)	2232	4718						
*Bgl*I	(2)	7472	8589						
*Bgl*II	(1)	4455							
*Bin*I	(21)	35	883	1220	2734	3752	3995	4699	4708
		4721	4810	4823	6176	6185	8068	8072	8389
		8852	8853	8949	8951	9037			
*Bsm*I	(4)	595	2673	2926	4055				
*Bsp*1286	(14)	428	1146	1236	1338	1399	2458	2469	5473
		5973	6977	7543	8039	8124	9285		
*Bsp*MI	(3)	936	1032	1537					
*Bsp*MII	(2)	2822	7006						
*Bst*EII	(1)	923							
*Bst*XI	(3)	749	908	3156					
*Bst*NI	(8)	332	487	977	7364	9436	9449	9570	9858
*Cla*I	(4)	251	4703	5857	6180				
*Dra*I	(14)	147	1484	1573	1835	2886	3418	4335	4975
		6300	6309	7171	8129	8821	8840		
*Dra*II	(4)	1395	1396	4569	7727				
*Eae*I	(10)	111	333	801	1504	1932	6473	7082	7328
		8314	9756						
*Eco*K	(2)	6328	8036						
*Eco*RI	(3)	1	4449	7310					
*Eco*RV	(1)	4441							

*Fok*I	(11)	865	1421	2168	2594	4300	4563	7382	7626
		8268	8555	8736					
*Fsp*I	(5)	2085	3648	6405	7462	8484			
*Hae*II	(12)	1733	4441	4845	5719	5884	5986	6028	6564
		6844	7485	9355	9725				
*Hga*I	(19)	824	922	1022	1388	1872	1925	1991	2455
		2693	3135	3492	3955	4945	5371	7128	7618
		8175	8905	9483					
*Hgia*I	(11)	428	1146	1236	1338	2458	2469	6977	7543
		8039	8124	9285					
*Hinc*II	(8)	1888	1915	2843	2949	3234	6254	7060	7190
*Hind*III	(2)	5248	6294						
*Hpa*I	(3)	2843	7060	7190					
*Hph*I	(15)	935	1160	1163	2598	3172	3868	5649	6655
		7688	7697	7980	8015	8221	8637	8864	
*Mbo*II	(20)	202	205	632	1243	1552	3057	3487	3490
		3493	3685	3781	5349	7119	7438	7925	8034
		8112	8867	8938	9729				
*Mlu*I	(3)	2469	3388	4663					
*Nae*I	(1)	3772							
*Nar*I	(4)	1730	5716	5983	7482				
*Ncl*I	(9)	407	4388	4573	4711	7636	7671	8171	8522
		9218							
*Nde*I	(1)	7534							
*Nsl*I	(3)	2673	3173	4590					
*Nsp*BII	(175)	870	1309	1360	1966	2639	3644	4163	4445
		5808	5988	6831	7412	7606	8071	9012	9257
		9775							
*Ppu*MI	(1)	4569							
*Pvu*I	(2)	7443	8338						
*Pvu*II	(5)	1309	4445	6831	7412	9775			
*Sac*II	(1)	871							
*Sal*I	(3)	2947	3232	6252					
*Sca*I	(1)	8226							
*Sfa*NI	(18)	1076	2381	2661	3161	4096	4566	4628	5649
		6396	6410	7497	7499	7573	7629	8025	8236
		8465	9517						
*Sna*BI	(1)	4885							
*Sph*I	(1)	2135							
*Ssp*I	(11)	274	587	2046	2996	3414	4979	5176	6049
		6305	6819	7902					
*Sty*I	(1)	1501							
*Xba*I	(1)	4694							
*Xho*I	(1)	1901							
*Xho*II	(9)	4455	4713	4815	8060	8077	8845	8857	8943
		8954							
*Xma*III	(1)	7082							
*Xmn*I	(4)	1247	5426	7010	8107				

[1] Possee *et al.* (1991); Weyer and Possee (1991); see also Figure 4.2.

Although the characterization of the transfer vectors with RE provides clear evidence of the orientation of the inserted fragment, it may be advantageous to consider sequence analysis of the junctions, especially at the 5′ end, between the insert and the transfer vector. This will provide absolute confirmation of the integrity of the construction. This may be readily accomplished by using custom oligonucleotide primers (with chain termination sequencing methods on double-stranded plasmid DNA) designed to read across the 5′ and 3′ ends of the inserted fragment. Figure 4.3 shows the sequence of an oligonucleotide primer that can be used to sequence across the 5′ cloning junction in pAcYM1.

Required:
- Mini-prep DNA samples (from section 4.7.2)
- Baculovirus transfer vector DNA (as a control)
- 10× RE buffer(s)
- REs, as required
- 5× DNA sample mix, for gel analysis (50 mM EDTA, 0.1% SDS, 0.1% bromophenol blue, 0.1% xylene cyanol FF, 50% sucrose)
- Materials for agarose gel electrophoresis
- 5 mg/ml ethidium bromide (take appropriate safety precautions)

Method:
1. Set up RE digests to (1) identify which of the mini-prep DNA clones has the required insert and (2) identify which of those with the required insert is in the correct orientation with respect to the polyhedrin gene. You may be able to achieve both objectives with a single digestion with one or two RE, or you may prefer to identify clones with inserts first and then undertake secondary digests on the positives to orientate the insert.
2. A typical digest might be:

Mini-prep plasmid DNA	0.5–2 μl
10× RE buffer	2 μl
1st RE (2 units)	0.2 μl
2nd RE (2 units) (if required)	0.2 μl
Sterile water	to 20 μl

   * The amount of plasmid DNA to be digested should be determined empirically using gel analysis of undigested DNA.

   * The single reaction mix described above may be prepared as a master mix (without the plasmid DNA) and then added to tubes containing the individual DNA samples. This greatly reduces the number of pipetting steps involved, when screening several mini-prep DNA samples.

   * The use of two RE in the same reaction buffer is dependent on

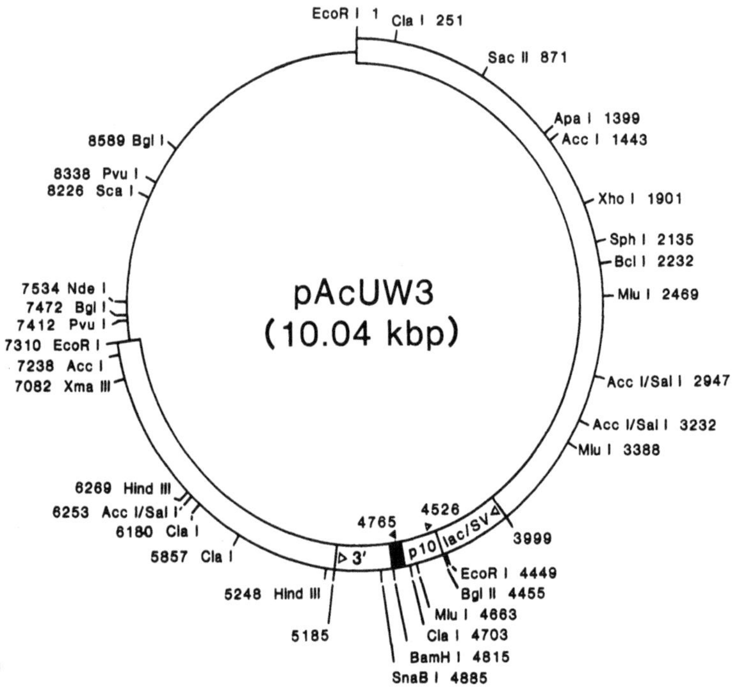

(a)

Figure 4.2 (a) Restriction enzyme map of the dual-expression vector pAcUW3 (Weyer and Possee, 1991). See Table 4.2 for a detailed listing of all the REs that recognize a 6-nucleotide sequence. The plasmid backbone (pUC derivative) is indicated by a single line and the AcMPNV sequences by a double line. The polyhedrin promoter is represented by a solid box with the transcription initiation site at position 4765 (solid triangle) and the 3′ end of the mRNA at position 5185 (open triangle). The p10 promoter is labelled and the transcription initiation site is shown at position 4526. Also indicated are the associated *lacZ* sequences (out of frame) and SV40 transcription termination signals (lacZ/SV). (b) Nucleotide sequence of the dual polyhedrin and p10 gene promoters in pAcUW3. The antisense strand of the p10 promoter is shown, with the transcription initiation site double underlined and labelled with a thick, horizontal arrow. The polyhedrin promoter is shown in the positive sense.

the required conditions for each enzyme. Consult the manufacturers' specifications if in doubt; many suppliers now provide a set of buffers to cover all requirements.

3. Digest 0.5–1.0 μg of the unmodified transfer vector with the same RE to serve as a control, for transfer vector with no inserted sequences.
4. Incubate the digests for 2 h at 37°C, then add 5 μl 5× DNA sample mix.

```
        4390      4400      4410      4420      4430      4440
GCCTGCCCGGTTATTATTATTTTTGACACCAGACCAACTGGTAATGGTAGCGACCGGCGC

        4450      4460      4470      4480      4490      4500
TCAGCTGGAATTCAGATCTGTGATTGTAAATAAAATGTAATTTACAGTATAGTATTTTAA
       EcoR I Bgl II
        4510      4520      4530      4540      4550      4560
TTAATATACAAATGATTTGATAATAATTCTTATTTAACTATAATATATTGTGTTGGGTTG
                            <-T   p 10
        4570      4580      4590      4600      4610      4620
AATTAAAGGTCCCGGCATCCTCAAATGCATAATTTCATAGTCCCCCTTGTTGTAAGTGAT

        4630      4640      4650      4660      4670      4680
GCGTATTTCTGAATCTTTGTAAAATAGCACACAAGACTCCAACGCGTTTGGCGTTTTATT
                                         Mlu I
        4690      4700      4710      4720      4730      4740
TTCTTGCTCGACTCTAGAGGATCGATCCCCCGGATCTGATCATGGAGATAATTAAAATGA
            Xba I   Cla I
        4750      4760 T->  4770      4780      4790      4800
TAACCATCTCGCAAATAAATAAGTATTTTACTGTTTTCGTAACAGTTTTGTAATAAAAAA
                         polyhedrin
        4810      4820      4830      4840      4850      4860
ACCTATAAATACGGATCCGGTTATTAGTACATTTATTAAGCGCTAGATTCTGTGCGTTGT
            BamH I
```

(b)

5. Analyse the digested DNA using an agarose minigel, with appropriate size markers (e.g. λ DNA digested with *Hin*dIII). Calculate DNA fragment sizes and thus determine whether the foreign DNA has been inserted and its orientation with respect to the virus promoter. A comparison with the control transfer vector DNA digest is invaluable here.

If it has not already been done, it is wise to store a little of the bacterial culture containing the transfer vector at −70°C in LB containing 15% glycerol. This is an insurance against possible mishaps when purifying the maxi-prep (see section 4.9)

4.9 ISOLATION OF HIGHLY PURIFIED PLASMID DNA (MAXI-PREPS)

There is still some debate over the necessity for baculovirus transfer vectors to be purified using the stringent procedures involving caesium chloride gradients. Some workers claim to have used mini-prep DNA directly in co-transfections with infectious virus DNA and produced recombinant virus without difficulty. We do recommend using a maxi-prep procedure, at least when using the system for the first time, to ensure successful co-transfections and recombinant virus production. A compromise may be to clean-up the mini-prep DNA using the

```
                Primer 5'-TGATAACCATCTCGCAA-3'
GATATCATGGAGATAATTAAAATGATAACCATCTCGCAAATAAATAAGTATTTTACTGTT
EcoR V (4000, pAcYM1)

TTCGTAACAGTTTTGTAATAAAAAAACCTATAAATACGGATCC-foreign gene coding
                                     BamH I (4095, pAcYM1)

sequence-GGTTATTAGTACATTTATTAAGCGCTAGATTCTGTGTGCGTTGTTGATTTA


CAGACAATTGTTGTACGTATTTTAATAATTCATTAAATTTATAATCTTTAGGGTGGTATG
                                                    3'-CACCATAC

TTAGAGCGAAAATCAAAT
AATCTCGCT-5' Primer
```

Figure 4.3 Nucleotide sequences of two oligonucleotide primers that can be used to sequence across the 5′ cloning junction (upper primer) in any polyhedrin promoter-based transfer vector and the 3′ cloning junction (lower primer). Note the 3′ primer is only applicable to transfer vectors where the polyhedrin coding region has been deleted e.g. pAcYM1.

recently developed affinity columns, e.g. those produced by Stratagene (PlasmidQuick) or Schleicher and Schuell (Elutip). The maxi-prep method method presented here will yield 0.5–1.0 mg DNA from a starting culture of 200 ml, with most of the baculovirus transfer vectors.

Required:

- LB with ampicillin (50 μg/ml, see section 4.6)
- Sucrose buffer (50 mM Tris-HCl, pH 8.0, 25% sucrose), autoclave at 10 p.s.i. for 10 min
- Tris/EDTA (0.25 M EDTA, 0.25 M Tris-HCl, pH 8.0)
- Triton lysis buffer (2% (v/v) Triton X-100, 50 mM Tris-HCl, pH 8.0, 10 mM EDTA), autoclave at 15 p.s.i. for 20 min
- Lysozyme (10 mg/ml in sucrose buffer), freshly prepared
- Ribonuclease A (10 mg/ml; see section 4.7.2)
- 75% and 100% ethanol
- 10% SDS
- Phenol/chloroform (50:50, equilibrated with 100 mM Tris-HCl, pH 8.0)
- 3 M sodium acetate
- TE buffer (10 mM Tris-HCl, pH 8.0, 1 mM EDTA)
- Caesium chloride
- Butanol
- Ethidium bromide (10 mg/ml)

Method:

1. Retrieve the appropriate 0.5 ml culture left over from the initial mini-prep screening procedure (section 4.7.2) and inoculate a 200

ml LB culture containing ampicillin. Shake vigorously at 37°C for 14–18 h.

* Ensure that the vessel is large enough to provide sufficient aeration of the bacterial culture (e.g. 250 ml in a 1000 ml flask).

2. Decant the culture to large centrifuge tubes (250 ml) and pellet the cells at 4000 r.p.m. (4°C).
3. Discard the supernatant fraction into a suitable disinfectant, resuspend the cells in 7.5 ml sucrose buffer and transfer to a disposable 50 ml polypropylene centrifuge tube.
4. Add 1.5 ml lysozyme, mix and stand on ice for 10 min.
5. Add 125 μl ribonuclease A and 3 ml Tris/EDTA solution. After mixing, stand on ice for 5 min.
6. Add 12.5 ml Triton lysis buffer and mix very gently. If the tube is placed on its side in the ice and slowly rotated, then the lysis of the cells may be monitored by the appearance of sticky strands which adhere to the sides of the tube. Stand on ice for 10 min.
7. Pellet the high molecular weight material at 20 000 r.p.m. in a Beckman SW28 (38 ml) rotor (or equivalent).
8. Remove the supernatant fraction to a polypropylene tube, add SDS to 1%, and extract with an equal volume of phenol/chloroform for 10 min.
9. Separate the phases in a low-speed bench centrifuge and transfer the aqueous phase to a glass or polypropylene tube.
10. Add sodium acetate to 0.3 M and two volumes of ethanol. Chill at −20°C for 30 min.
11. Pellet the DNA at 3000 r.p.m. (low-speed laboratory centrifuge) and wash with 75% ethanol. Drain the excess ethanol, briefly air-dry the pellet and then dissolve in 7.5 ml TE buffer.
12. Add 8 g caesium chloride, dissolve, and then add ethidium bromide to 100 μg/ml.
13. Add the DNA/caesium chloride mixture to a 16 × 76 mm sealable ultracentrifuge tube (e.g. Beckman quick seal) and top up with liquid paraffin. Prepare a balance tube if only one sample is being processed. Seal the tubes and place in a 70.1 Ti Beckman rotor (or equivalent) and spin for 18–24 h at 45 000 r.p.m.
14. Harvest the plasmid by puncturing the side of the tube with a large-gauge needle attached to a 2–5 ml syringe and carefully draw off the lower DNA (plasmid) band.

    * The use of a second gradient at this stage results in a considerable improvement in the purity of the plasmid DNA isolated with this method, but is not necessary for co-transfection experiments.
15. Remove the ethidium bromide dye by extraction with butanol. Repeat the extractions until the butanol remains colourless.

16. Dilute the DNA by adding 2 volumes TE buffer, add sodium acetate to 0.3 M, then add two volumes of ethanol. There is no need to chill the DNA at this stage.
17. Pellet the DNA using a microfuge or low-speed laboratory centrifuge. After washing the DNA pellet once with 75% ethanol, resuspend in 1 ml TE buffer and add SDS to 1%.
18. Transfer the DNA solution between two microfuge tubes, extract with an equal volume of phenol/chloroform for 5 min at room temperature and spin in a microfuge to separate the phases.
19. Transfer the aqueous phase into fresh tubes and ethanol precipitate the DNA with sodium acetate and ethanol, as described above (16).
20. Wash the DNA twice with 75% ethanol and dry the pellet. Resuspend the DNA in 500 μl TE buffer and store at 4°C or −20°C.

    * The above procedure is slower than other plasmid purification methods but does yield very high quality DNA; particularly if a second caesium chloride gradient is used.

Insect cell culture media and maintenance of insect cell lines 5

5.1 INTRODUCTION

This chapter covers all aspects of insect cell culture that are required to obtain success with the baculovirus expression vector system. After the introductory comments, detailed protocols for the preparation and storage of culture media, routine maintenance of insect cells in culture and the long-term preservation of insect cells in liquid nitrogen are presented.

5.2 CELL LINES

Cell lines derived from *S. frugiperda*, the fall army worm, are used for most of the expression studies with AcMNPV (reviewed by Cameron *et al.*, 1989). The original and still widely used cell line is Sf21 (IPLB-Sf-21; Vaughn *et al.*, 1977) which was derived from ovarian tissue. The alternative line, Sf9 (Smith *et al.*, 1983a), is a clonal derivative of Sf21 and is also very widely used. Both cell lines can be passaged continuously and in our labs Sf21 cells are routinely used between passage numbers 150 and 180. This is a convention that we have adopted to ensure reasonable reproducibility of results after virus infections. Frozen stocks are kept of the lower passage number cells and these are used for a period of about six months after thawing. Sf21 cells are available on request from Oxford and Sf9 cells are available from the American Type Culture Collection, Accession Number CRL-1711 (addresses in Appendix A). There are some unpublished observations that Sf9 and Sf21 cells yield different levels and process recombinant proteins in slightly different ways and, therefore, it may be worthwhile testing expression of a new gene in both cell lines. In particular we strongly recommend that expression of receptors for electrophysiological analyses be carried out in Sf9 cells (King *et al.*, in press). An alternative cell line, derived from *Mamestra brassicae* (Mb) appears to provide an increase in both polyhedrin and foreign protein levels that are about two- or three-fold higher than are given with either Sf21 or Sf9 cells. Unfortunately, these cells are not as easy to use as the Sf lines (King *et al.*, 1991a; cells available on request). It is therefore recommended that the Mb cells are only tested if low levels of expression are obtained using the Sf cells.

5.3 CULTURE MEDIA

The Sf21, Sf9 and Mb cells can be maintained in TC100 or Grace's medium (Grace, 1962). Both media need to be supplemented with foetal calf serum (FCS) and, in addition, Grace's basal medium must be supplemented with yeastolate and lactalbumin hydrolysate; when supplemented in this manner, Grace's medium is often referred to as TNM-FH medium (Hink, 1970). Media can also be supplemented with antibiotics, if desired. More recently, other media have been developed for insect cell culture and some of these do not require supplementing with FCS. For example, SF900 and IPL-41 media from Gibco Life Technologies (LT) and EX-CELL 400/401 medium from J.R. Scientific or Sera Lab can be used serum-free. Cells growing in serum-containing media need to be weaned onto the serum-free media gradually. Insect cell culture media that are available commercially are listed in Table 5.1. TC100, Grace's, SF900, IPL-41 and EX-CELL 400/401 media can be purchased in ready-to-use liquid form. For economy, TC100, IPL-41 and EX-CELL 401 can be obtained in a powder form for the simple preparation of liquid media. TC100 can also be made up from individual ingredients, and although this is not recommended for routine use, it will probably be necessary when preparing specialized media, such as TC100-minus methionine, for metabolic-radiolabelling experiments. Insect cell culture media do not contain pH indicators, hence their pale yellow colour. The pH when formulated is 6.18–6.2, a value which will rise when the cells are growing but which should not usually exceed 6.4.

In this chapter we give details for preparing TC100 and Grace's (TNM-FH) media for routine use and TC100 for specialized uses, such as metabolic labelling studies. In the experimental chapters that follow (Chapters 6–8), for simplicity we state the use of TC100 medium. You can, of course, substitute Grace's (TNM-FH) medium or one of the newer media mentioned above in any of these protocols.

Although many papers recommend the use of FCS at 10%, we have found that 5% FCS has no adverse effects on the cells, apart from a slightly reduced rate of growth; the titre of virus stocks is unaffected.

Table 5.1 Insect cell culture media and supplements that are available commercially

Medium or supplement	*Supplier*	*Liquid (L) Powder (P)*	*Comments (purchase number)*
Media			
TC100	Gibco LT	1× L	500 ml, 041-03055M
		P	1–50 l, 074-03000
	ICN-Flow	1× L	450 ml, 73-715-53
	Sigma[1]	P	1–50 l, T0907
	Sera Lab[2]	P	10 l, 56-941-110
Grace's medium without supplements	Gibco LT	1× L	500 ml, 041-01590M
	Sigma	P	1–50 l, G9771
	Sera Lab[2]	P	10 l, 56-902-110
Grace's medium with supplements (TNM-FH)	Gibco LT	1× L	500 ml, 041-01605M
	Sera Lab[2]	1× L	500 ml, 51-94278
	Sigma	P	1–50 l, T1032
IPL-41	Gibco LT	1× L	500 ml, 041-01405M
		P	1–50 l, 074-04000
	Sigma	P	1–50 l, I0638
	Sera Lab[2]	1× L	500 ml, 51-92478
		P	10 l, 56-923-110
IPL-41 (minus methionine)		P	10 l, 56-925-110
SF900	Gibco LT	1× L	500 ml, 041-00900M
EX-CELL 400[3]	Sera Lab[2]	1× L	500 ml, 14-40078
EX-CELL 401[3]		1× L	500 ml, 14-40178
EX-CELL 401[3] (minus methionine)		1× L	500 ml, 14-40378
EX-CELL 401[3] (minus phosphate)		1× L	500 ml, 14-40478
EX-CELL 401[3]		P	10 l, 14-410-110
Serum			
Foetal calf serum (FCS)	Gibco LT	100/500 ml	011-06290H/M
	Sigma	100/500 ml	F0643
	GlobePharm	100/500 ml	–
	Sera Lab[2]	500 ml	S0001a
	ICN-Flow	500 ml	29-101-54
Heat-inactivated FCS	Gibco LT	100/500 ml	013-06290H/M
	Sigma	100/500 ml	F3018
	GlobePharm	100/500 ml	–
	Sera Lab[2]	500 ml	HI0001a
	ICN-Flow	500 ml	29-102-54
Supplements			
Lactalbumin hydrolysate (Grace's medium)	Gibco LT	50× L	100 ml, 043-08080H
	Sigma	P	250 g, L4896
	Sera Lab[2]	50× L	100 ml, 58-90177
Yeastolate (Grace's medium)	Gibco LT	50× L	100 ml, 043-08190H

Medium or supplement	*Supplier*	*Liquid (L) Powder (P)*	*Comments (purchase number)*
	Sera Lab[2]	50× L	100 ml, 58-90277
	Sigma	P	100 g, Y1000
Yeastolate (IPL-41)	Gibco LT	50× L	100 ml, 043-08200H
	Sera Lab[2]	50× L	as for Grace's
	Sigma	P	as for Grace's
Tryptose phosphate broth (IPL-41)	Gibco LT	50× L	100 ml, 043-08061H
	ICN-Flow	50× L	100 ml, 16-821-49
	Sigma	P	100 g, T4532
	Sera Lab[2]	50× L	100 ml, 58-90077
Lipid mixture (IPL-41)	Gibco LT	100× L	100 ml, 066-01900H
	Sigma	100× L	100 ml, L5146
Pluronic F-68[4]	Gibco LT	10% solution	100 ml, 066-04040H
	Sera Lab[2]	10% solution	100 ml, 59-91577
	Sigma	–	100 g, P1300
Tween 80	Sigma	L	100 ml, P4675
D-α-tocopherol acetate	Sigma	–	1 g, T1157
Cod liver oil	Sigma	–	1 g, C2294
Cholesterol	Sigma	–	1 g C2044
Antibiotics			
Gentamicin (10 mg/ml)	Gibco LT	20/100 ml	043-05710D/H
	ICN-Flow	10 ml	16-760-45
	Sigma	10 ml	G1272
Penicillin-streptomycin (5000)	Gibco LT	20/100 ml	043-05070D/H
	ICN-Flow	100 ml	16-700-49
	Sigma	20/100 ml	P0906
Nystatin	Gibco LT	20/100 ml	043-05340D/H
	Sigma	20 ml	N1638
Antibiotic/antimycotic mix	Gibco LT	20–100 ml	061-05240D/H
	Sigma	20/100 ml	A9909
Miscellaneous			
Phosphate buffered saline	Gibco LT	10× L	500 ml, 042-04200M
		P	1–50 l, 076-01300
	ICN-Flow	tablets	100, 28-103-05
	Sigma	P	1–10 l, D5652
Neutral red	Sigma	P	1 g, N4638
	ICN-Flow	L	100 ml, 16-911-49
Trypan blue	Sigma	P	5 g, T6146
	ICN-Flow	L	100 ml, 16-910-49

[1] The recipe for TC100 used by Sigma is slightly different from that used by the other manufacturers and that given in this book (see section 5.4.4).
[2] Use J.R. Scientific in the USA.
[3] A Trademark of J.R. Scientific.
[4] A Trademark of BASF Corporation.

One procedure where the use of 10% FCS is recommended, however, is the freezing of cells for storage in liquid nitrogen. You can purchase FCS from a number of suppliers (see Table 5.1). We have successfully weaned Sf21 and Sf9 cells onto SF900, EX-CELL 400 and IPL-41 media, but the cost benefits of using such media compared with using serum at 5% are minimal (when purchasing ready-to-use liquid media). However, there are certainly advantages in using a serum-free medium when trying to assay or purify secreted foreign proteins.

Whether or not routinely to include antibiotics in media is a question of personal choice. Some groups prefer to maintain cell lines in an antibiotic-free medium and only add antibiotics when carrying out virus infections, for example, plaque-assays. It is possible that by including antibiotics routinely, a low-level infection is maintained in the cells. Several different antibiotics may be chosen but the most widely quoted are penicillin–streptomycin and/or gentamicin. Some workers also add a fungicide such as nystatin. Suppliers of antibiotics are given in Table 5.1.

5.4 PREPARATION OF CULTURE MEDIA

This section assumes that, at least initially, you will be using either TC100 or Grace's (TNM-FM) medium supplemented with FCS. Guidelines for the preparation and use of alternative media (SF900, IPL- 41 and EX-CELL 400/401) are given in section 5.4.6.

5.4.1 PREPARATION OF TC100/FCS GROWTH MEDIUM

In this book media containing FCS that are ready-for-use will be referred to as growth media or TC100/FCS, for example. Whether you use FCS at 10% or 5% is your decision. Where medium supplemented with 10% FCS is required this will be clearly stated.

Required:

1 bottle (500 ml) TC100 (purchased or prepared as in sections 5.4.3 or 5.4.4)
Foetal calf serum (see notes below)
Antibiotics, if using (see notes below)

Method:

1. Remove a bottle of TC100 (commercially formulated or as prepared in sections 5.4.3 or 5.4.4) from storage at 4°C and wipe the bottle with 70% alcohol prior to placing in the culture cabinet.
   * Commercially formulated (1× liquid) TC100 can be purchased from the suppliers listed in Table 5.1. It is already supplemented with sodium bicarbonate.

2. Using aseptic technique, add 25 ml (for 5% v/v) or 50 ml (for 10% v/v) heat-inactivated FCS to 500 ml TC100 (450 ml if prepared as described in sections 5.4.3 or 5.4.4).
   * Recommended suppliers of FCS and heat-inactivated (HI) FCS are given in Table 5.1. Heat-inactivated FCS may be prepared in the laboratory using the following procedure: using aseptic technique, dispense FCS into convenient size aliquots in sterile glass bottles or plastic tissue culture tubes. Place in a VERY CLEAN water bath at 56°C and, after allowing time for the FCS to warm to 56°C, heat for 30 min. Wipe tubes/bottles with 70% alcohol and store frozen at −20°C.
3. If required, aseptically add antibiotics and fungicide, according to the manufacturers' recommended levels. For example, add 5 ml of penicillin–streptomycin (5000 units/ml).
   * Recommended antibiotics and fungicides are listed in Table 5.1.
4. TC100 supplemented with FCS (growth medium) should be stored at 4°C and has a shelf-life of 4–6 weeks.
5. If preferred, a sterility check can be made before using the TC100 growth medium (see section 5.4.3, point 8, for details).

5.4.2 PREPARATION OF GRACE'S (TNM-FH) GROWTH MEDIUM

Grace's medium as originally formulated (referred to here as basal medium) must be supplemented with yeastolate and lactalbumin hydrolysate for the culture of Sf cells. After addition of these supplements the medium is usually referred to as TNM-FH medium.

Required:
1 bottle (500 ml) Grace's medium (with or without supplements; see Table 5.1)
FCS (HI) (see section 5.4.1 for more details)
Antibiotics, if using
*Yeastolate supplement (see point 2 below)
*Lactalbumin hydrolysate supplement (see point 2 below)

Method:
1. Remove a bottle of Grace's medium (commercially formulated) from storage at 4°C and wipe the bottle with 70% alcohol prior to placing in the culture cabinet.
2. If using the basal medium, add the supplements as follows. To 500 ml medium add 10 ml 50× yeastolate and 10 ml 50× lactalbumin hydrolysate.

* The supplements are most easily obtained as 'ready-for-use' reagents from a commercial supplier, as listed in Table 5.1.

3. Using aseptic technique, add 25 ml (for 5% v/v) or 50 ml (for 10% v/v) HI-FCS to 500 ml Grace's (TMN-FH) medium (see section 5.4.1 for heat-inactivation procedure).
4. If required, aseptically add antibiotics and fungicide, according to the manufacturers' recommended levels, as described in section 5.4.1.
5. Grace's (TNM-FH) medium supplemented with FCS (growth medium) should be stored at 4°C and has a shelf-life of 4–6 weeks.
6. If preferred, a sterility check can be made before using the growth medium (see section 5.4.3, point 8, for details).

5.4.3 PREPARATION OF 4.5 l TC100 MEDIUM (APPROX. 5 l INCLUDING FCS) FROM POWDERED FORMULA

This is probably the cheapest medium for routine use and involves only a minimum amount of extra effort.

Required:

TC100 powder (see Table 5.1 for suppliers)
5 l pyrex flask (mark a 4.5 l guideline)
$NaHCO_3$
Distilled and/or deionized water
2 M HCl
10 M KOH
Supply of sterile 500 ml bottles
Filtering system (see notes below)
Accurate pH meter
Stirrer and stirring bar

* Use the best quality water available
* Keep the medium stirring throughout the preparation stages.
* Check the pH meter is accurate with standard buffers before starting
* See section 5.5 for the preparation and cleaning of glassware for tissue-culture use
* The filtering stage should be undertaken in a laminar flow hood.

Method:

1. Weigh sufficient powder for 5 l (final volume) of TC100 (99.7 g) and add to 4 l water. Stir until all the lumps have disappeared, although the medium will still look cloudy at this stage.

2. Add 45 ml 2 M HCl; the medium should clear. Stir for 10–15 min.
3. Dissolve 1.75 g $NaHCO_3$ in 200 ml water and then add to the medium
4. Make volume up to 4.5 l with distilled/deionized water.
5. Adjust the pH to 6.18 by adding 10 M KOH in small aliquots (1–2 ml at a time). After addition of each aliquot a precipitate will form which should clear in 20–30 s. It is critical that the pH does not rise above pH 6.3; above pH 6.3 the precipitate formed is insoluble (and you have to start again).
6. After adjusting the pH, stir for a further 5–10 min.
7. Filter-sterilize the medium into sterile bottles.
   * Working in a laminar flow or Class II safety cabinet, filter the medium into sterile glass bottles; 450 ml TC100 into 500 ml bottles is most convenient. Use either a positive pressure vessel system, if available, through a specially tested 0.2 μm cell culture filter with bell-jar protective filling-hood (e.g. Millipore or Gelman Sciences) or use a disposable bottle filtering system. Convenient 500 ml and 100 ml bottle filters can be purchased (e.g. Costar from NBL Ltd) and are very economical. Attach a bottle filter to the neck of the first bottle and attach the side-arm to a water vacuum line (with vacuum-trap in place). Place a pre-filter (supplied) in position and add 450 ml TC100 medium, turn on the vacuum (ensure pre-filter is in place with blunt-ended forceps). The 450 ml TC100 should filter through in 1–2 min. Release vacuum and transfer bottle filter aseptically to the second bottle; cap the first bottle. Continue until all 10 bottles are completed. When the filtration rate becomes very slow (after 5–10 bottles), use a new pre-filter or new bottle filter, as appropriate.
8. Set up sterility checks by sampling at least the first and last bottle of medium prepared from each bottle filter used. Place 1–2 ml aliquots at 28°C and 37°C for 3 days prior to using the medium (you can also add 1–2 ml samples to 10 ml bacterial broth and incubate at 37°C for 2 days). If the samples are clear after incubation, the TC100 is ready for use.
9. Cover the bottles with foil to prevent dust settling onto the caps and store at 4°C. TC100 kept in this way has a shelf-life of at least 1 year.
10. Prepare TC100 growth medium as described in section 5.4.1.

5.4.4 PREPARATION OF TC100 MEDIUM FROM INDIVIDUAL INGREDIENTS

This is the cheapest way of preparing TC100 medium, if labour costs are excluded. It is not recommended that TC100 for routine use be prepared in this way, unless commercial supplies are not available for some reason. Variations of TC100 medium, for example, lacking amino acids or sugars for metabolic-radiolabelling studies will, however, have to be prepared in this manner as commercial supplies are not generally available; Gibco LT will, however, make such media as specialized orders. Because only small quantities of these types of TC100 are normally required, recipes for 900 ml TC100 as well as 4.5 l are given. The 900 ml volume is more conveniently stored in 90 ml aliquots.

Great care must be taken to add ingredients in the order stated or else problems with insoluble precipitates will almost certainly occur. We recommend that you read through the check-list (Table 5.2), if not the detailed method, before starting.

In the following protocol details for preparing 900 ml TC100 are given in square brackets, after the quantities for 4.5 l. If you require specialized media for metabolic labelling experiments refer to Table 5.3 to see which ingredients to omit or modify; the remaining method is exactly the same as given below for complete TC100.

Required:
- 5 l [1 l] flask (marked at the 4.5 l [900 ml] level)
- Magnetic stirrer and bar
- Accurate pH meter (check with standard buffers before starting)
- Sterile 500 ml [100 ml] bottles
- Filtering system (see comments in section 5.4.3)
- 2 M HCl
- 10 M and 1 M KOH
- 10% (v/v) and concentrated glacial acetic acid
- Vitamins as detailed in 1a
- Soluble amino acids as detailed in 2a
- Insoluble amino acids as detailed in 2b
- Bactotryptose Broth (Difco Laboratories Ltd)
- Glucose (Sigma)
- KCl
- $CaCl_2.H_2O$
- $MgCl_2.6H_2O$
- $MgSO_4.7H_2O$
- $NaH_2PO_4.2H_2O$
- $NaHCO_3$

Table 5.2 Check-list of ingredients for the preparation of TC100 medium

Vitamin mixture prepared in advance:

	g/l
Thiamine HCl	0.02
D-Pantothenic acid-Ca^{2+}	0.11
Pyridoxine-HCl	0.02
Nicotinic acid	0.02
Iso-(myo-)inositol	0.02
Cyanocobalamin	0.01
p-Aminobenzoic acid	0.02
Folic acid	0.02
Riboflavin	0.02
D-Biotin	0.01

Complete TC100 media:

	4.5 l batch	900 ml batch
Vitamin mixture (prepared as above)	5 ml	1 ml
Soluble amino acids:		
L-Alanine	1.13 g	0.225 g
L-Arginine	2.75	0.55
L-Asparagine	1.75	0.35
L-Aspartic acid	1.75	0.35
L-Glutamic acid	3.00	0.60
L-Glutamine	3.00	0.60
L-Glycine	3.25	0.63
L-Histidine	16.90	3.40
L-Isoleucine	0.25	0.05
L-Leucine	0.38	0.075
L-Lysine	3.13	0.63
L-Methionine	0.25	0.05
L-Phenylalanine	0.75	0.15
L-Proline	1.75	0.35
L-Serine	2.75	0.55
L-Threonine	0.90	0.18
L-Valine	0.50	0.10
Insoluble amino acids:		
L-Cysteine	0.11 g	0.02 g
L-Tryptophan	0.50	0.10
L-Tyrosine	0.25	0.05
Other ingredients:		
Bactotryptose broth	13.0 g	2.6 g
KCl	14.35	2.87
$CaCl_2.H_2O$	7.6	1.32
Glucose	5.0	1.0
$MgCl_2.6H_2O$	11.4	2.28
$MgSO_4.7H_2O$	13.9	2.78
$NaH_2PO_4.2H_2O$	5.7	1.14
$NaHCO_3$	1.75	0.35

Table 5.3 Specialized TC100 media

Media	*Purpose*	*Modification*
TC100 minus methionine	Protein labelling with [^{35}S]methionine	Omit Methionine[1] and bactotryptose[1]
TC100 minus phosphate	[^{32}P]phosphate labelling	Omit $NaHPO_4.2H_2O$
TC100/10% glucose	Glycosylation experiments using labelled sugars	Use 10% stated amount of glucose
TC100 minus arginine	Labelling basic proteins with [^{14}C]arginine	Omit arginine[1] and bactotryptose
TC100 minus amino acid acid of choice	Protein labelling with relevant amino acid	Omit relevant amino acid[1] and bactotryptose

[1] Omit from relevant stock amino acid solution.

* The vitamin solution is prepared in batch amounts and stored frozen at −20°C.
* All salts should be of Analar quality and must completely dissolved, where indicated, before adding to the other ingredients. It is best to weigh all salts before starting to prepare the media.
* A check-list of ingredients for complete TC100 medium will be found in Table 5.2.

Method:

1. Prepare 1 l of the stock vitamin solution. Weigh the following and dissolve together in 600 ml water, except for those vitamins numbered (1) or (2) which must be dissolved separately, as indicated.

	g/l
Thiamine-HCl (Sigma T1270)	0.02
D-Pantothenic acid (calcium salt; Sigma P5155)	0.11
Pyridoxine-HCl (Sigma P6280)	0.02
Nicotinic acid (Niacin; Sigma N0761)	0.02
Iso-(myo-)inositol (Sigma I7508)	0.02
Cyanocobalamin (Vitamin B12; Sigma V6629)	0.01
p-Aminobenzoic acid (Sigma A3659)	0.02 (1)
Folic acid (Sigma F8758)	0.02 (2)
Riboflavin (Sigma R9504)	0.02 (2)
D-Biotin (Sigma B4639)	0.01 (2)

Notes: (1) Dissolve separately in 25 ml 1 M KOH.
(2) Dissolve each separately in 25 ml 10% v/v glacial acetic acid (a drop of concentrated glacial acetic acid may also need to be added).

Add the separately dissolved vitamins to the water-soluble vitamins and increase the volume to 1 l. Dispense into 5 ml or 100 ml volumes and store at −20°C. There is no requirement to filter sterilize at this point.

For 4.5 l [900 ml] TC100 you will need to thaw out 5 ml [1 ml] vitamin solution.

2a. Prepare the soluble amino acid solution. Weigh and combine all the following amino acids in a mortar, then grind thoroughly with a pestle before dissolving in 1 l [200 ml] water. It is important to stir the amino acid solution for at least 30 min before adding remaining ingredients.
* Remember to omit the necessary amino acid if preparing specialized TC100 for metabolic labelling studies (see Table 5.3)

	g/4.5 l Batch	g/900 ml Batch
L-Alanine (Sigma A3534)	1.13	0.225
L-Arginine (Sigma A3784)	2.75	0.55
L-Asparagine (Sigma A4159)	1.75	0.35
L-Aspartic acid (Sigma A4534)	1.75	0.35
L-Glutamic acid (Sigma G5638)	3.00	0.60
L-Glutamine (Sigma G5763)	3.00	0.60
L-Glycine (Sigma G6388)	3.25	0.63
L-Histidine-HCl (Sigma H9511)	16.90	3.40
L-Isoleucine (Sigma I7383)	0.25	0.05
L-Leucine (Sigma L1512)	0.38	0.075
L-Lysine-HCl (Sigma L1262)	3.13	0.63
L-Methionine (Sigma M2893)	0.25	0.05
L-Phenylalanine (Sigma P5030)	0.75	0.15
L-Proline (Sigma P4655)	1.75	0.35
L-Serine (Sigma S5511)	2.75	0.55
L-Threonine (Sigma T1645)	0.90	0.18
L-Valine (Sigma V6504)	0.50	0.10

2b. Prepare the insoluble amino acid solution. Weigh the following amino acids, combine, and dissolve in 475 ml [95 ml] water and 25 ml [5 ml] 2 M HCl. Stir until dissolved; warming to 40°C may help.
* Remember to omit the necessary amino acid when preparing specialized TC100 for metabolic labelling studies (see Table 5.3).

	g/4.5 l Batch	g/900 ml Batch
L-Cysteine (Sigma S8152)	0.11	0.02
L-Tryptophan (Sigma T0271)	0.50	0.10
L-Tyrosine (Sigma T1020)	0.25	0.05

3. To 1 l [200 ml] dissolved soluble amino acids add 5 ml [1ml] vitamin solution.
   * From this point on keep the medium stirring at all times.
4. Dissolve 13 g [2.6 g] bactotryptose broth in 250 ml [50 ml] water and add to the amino acid/vitamin solution.
5. Dissolve 14.35 g [2.87 g] KCl, 7.6 g [1.32 g] $CaCl_2.H_20$ and 5 g [1 g] glucose in 400 ml [80 ml] water and add to medium.
6. Dissolve 11.4 g [2.28 g] $MgCl_2.6H_2O$ in 250 ml [50 ml] water and add to medium.
7. Dissolve 13.9 g [2.78 g] $MgSO_4.7H_2O$ in 250 ml [50 ml] water and add to medium.
8. Adjust the pH of the medium to 6.2 with 2 M HCl.
9. Dissolve 5.7 g [1.14 g] $NaH_2PO_4.2H_2O$ and 1.75 g [0.35 g] $NaHCO_3$ in 250 ml [50 ml] water and add to medium.
10. Add the 500 ml [100 ml] insoluble amino acid solution to the medium.
11. Make up volume to 4.5 l [900 ml] with water and adjust pH to 6.18 very slowly with 10 M KOH (stirring rapidly but without frothing). A precipitate will form and dissolve within 20–30 seconds upon each addition of KOH, however, if the pH rises above pH 6.3 an insoluble precipitate forms (and you will have to start again).
12. Filter-sterilize, sterility check and store medium as described in section 5.4.3. The 900 ml amount can be dispensed into 2 × 450 ml or in 10 × 90 ml amounts using 500 or 100 ml disposable bottle filters, as appropriate.

5.4.5 SPECIALIZED TC100 MEDIA

Make up 900 ml amounts of TC100 medium following the protocol in section 5.4.4, remembering to omit or modify the required ingredients, as indicated in Table 5.3. It is convenient to filter sterilize the media in 90 ml amounts. Store at 4°C.

5.4.6 ALTERNATIVE INSECT CELL CULTURE MEDIA

The following section briefly outlines other types of media that are available commercially for the culture of Sf cells and propagation of wild-type and recombinant baculoviruses.

(a) SF900 medium

This medium was developed by and is available from Gibco LT (see Table 5.1) and is a complete, ready-to-use, serum-free medium

(Weiss *et al.*, 1990). No supplements need to be added. Cells need to be weaned very carefully onto this medium (see section 5.9) or alternatively, Sf9 cells already adapted to SF900 may be obtained (free-of-charge) by writing to Gibco LT.

(b) IPL-41 medium

This medium was designed to optimize the growth of Sf9 cells in large-scale spinner cultures (Weiss *et al.*, 1981; Weiss and Vaughn, 1986; Maiorella *et al.*, 1988). It can be purchased in 1× liquid form or in powder form (see Table 5.1). The medium can be supplemented with FCS and tryptose phosphate broth, or it can be used as a serum-free medium when supplemented with yeastolate and a lipid concentrate. In the latter case, cells will need to be weaned onto the serum-free version (see section 5.9).

To prepare IPL-41 for use with FCS:

Required:
- IPL-41 liquid medium (purchased or prepared from powder as described below, point 3)
- HI-FCS (see section 5.4.1 for more details)
- Tryptose phosphate broth supplement (see note below)
- 1 M HCl
- 2.5 M NaOH
- $NaHCO_3$ (if using powdered formula)
- Sterile bottles and filtering system (if using powdered formula; see section 5.4.3 for more details)

Method:
1. Using aseptic technique, add 10 ml 50× tryptose phosphate broth to 500 ml 1× liquid IPL-41 (2.6 g/l final concentration).
   * 50× tryptose phosphate broth can be obtained commercially (see Table 5.1) or can be prepared by dissolving 13 g tryptose phosphate broth (Sigma) in 100 ml water (this may require warming to aid dissolution), adjusting the pH to 6.2 with 1 M HCl and filter sterilizing. Store sterile concentrate at 4°C.
2. Add HI-FCS to 5% or 10%, as desired.
3. To prepare 1× liquid IPL-41 from powdered formula:
   add required amount of powder to a volume of water 10–15% less than the final volume required. Stir at room temperature and add 0.35 g/l (final volume) NaHCO3. Adjust to final volume, pH to 6.1–6.2 with 2.5 M NaOH and filter sterilize.

To prepare IPL-41 for use as a serum-free medium:

Required:
- IPL-41 1× liquid medium (purchased or prepared from powder as described below, point 2)
- Yeastolate supplement (see below)
- Lipid concentrate supplement (see below)
- 2.5 M NaOH
- 15% (w/v) NaCl
- $NaHCO_3$ (if using powdered formula)
- Sterile bottles and filtering system (see section 5.4.3 for more details)

Method:
1. To 450 ml IPL-41 1× liquid medium, add 10 ml 50× yeastolate solution and 5 ml of 100× lipid concentrate (pre-warmed to 37°C). Adjust pH to between 6.1 and 6.2 with 2.5 M NaOH and osmolarity to 355–365 mOsm/kg with 15% (w/v) NaCl. Adjust final volume to 500 ml with water and filter sterilize. Store at 4°C.
   * 50× yeastolate can be obtained commercially (see Table 5.1) or can be prepared by dissolving 4 g/l yeastolate (Sigma or Difco) in water and passing through a hollow-fibre ultrafiltration system (cut-off point 10 000 molecular weight).
   * 100× lipid concentrate can be purchased commercially (see Table 5.1) or can be prepared by mixing the following together: cholesterol, 0.45 g/l; cod liver oil, 1 g/l; pluronic F-68, 100 g/l; D-alpha-tocopherol acetate, 0.2 g/l; Tween 80, 2.5 g/l. Store at room temperature but warm to 37°C before adding to the medium. All lipids can be obtained from Sigma (see Table 5.1).
2. To prepare 1× liquid IPL-41 from powdered formula: add required amount of powder to a volume of water 10–15% less than the final volume required. Stir at room temperature and add 0.35 g/l (final volume) $NaHCO_3$. Adjust volume to 10% less than the final volume, add supplements as described in point 1 and filter sterilize.

(c) EX-CELL 400/401

These media are available from J.R. Scientific (USA) or Sera Lab (Europe) in 1× liquid form and can be used without further supplementation. As with the SF900 medium, cells will need to be adapted to the media over a period of time (see section 5.9). EX-CELL 401 is a modified and updated version of 400 that is also available in powdered form and as 1× liquid medium deficient in methionine or phosphate, for metabolic labelling studies (see Table 5.1).

(d) Pluronic F-68

Pluronic F-68 (a trade mark of BASF Corporation) is a surfactant that has been shown to be effective in reducing the shear stress of cells (mammalian, hybridoma and insect) grown in spinners, shakers and in bio-reactors (Murhammer and Goochee, 1988; Tramper *et al.*, 1986). It is added routinely to IPL-41 and EX-CELL 400/401 media that are used serum-free (see above) but can also be used with other insect cell culture media at a 0.05–0.1% (v/v) concentration.

5.5 GLASSWARE AND DISPOSABLE PLASTICWARE

All plastic flasks, Petri dishes and pipettes need to be of cell culture quality and can be purchased from a number of suppliers including Gibco-Nunc, ICN-Flow, Sterilin, Sigma, Costar from NBL, etc. Routine maintenance of monolayer cultures can be carried out more cheaply if reusable glass flasks (e.g. ICN-Flow) are used. Re-usable glass pipettes can also be employed, but they must be thoroughly cleaned and rinsed after use, and preferably acid washed on each occasion.

You should adopt a rigorous routine for cleaning and rinsing all glassware used for cell work (flasks, bottles for storing media, scrapers, etc.). The regime that we use is described below. If possible, keep cell culture glassware separate from other laboratory glassware.

5.5.1 SUGGESTED CLEANING REGIME FOR TISSUE CULTURE GLASSWARE

Required:

- 5% w/v SDS; standard grade
- Lipsol or other laboratory detergent
- Hypochlorite or other disinfectant
- Distilled/deionized water
- Facilities for acid-washing glass pipettes (if using)

Method:

1. After use, glass culture flasks should be rinsed with 5% w/v SDS solution to remove any remaining cells and clean the glassware. Re-usable cell scrapers should be soaked in a dilute detergent solution until ready to be washed. Empty media bottles can be rinsed with dilute detergent or 5% w/v SDS.
   * If culture flasks become contaminated with bacteria/fungi, try to avoid autoclaving before cleaning, as this tends to bake cells

onto the flasks which are then very difficult to clean. Instead, disinfect flasks by soaking in hypochlorite and then rinse with 5% SDS solution as above.

* If cells become accidently infected with baculovirus, you should autoclave the flasks before washing, as polyhedra can be very resilient to chemical disinfection.

2. After cleaning in detergent or 5% SDS, glassware should be rinsed well with tap water and left standing in tap water until the proper rinsing regime can be started.
3. Rinse all glassware (including bottle tops and scrapers) ten times with tap water and twice with distilled water. Soak overnight and then give a final rinse in distilled water.
4. Dry bottles in warm air. Autoclave to sterilize and, if necessary, dry bottles before using for media or cells. Scrapers can be wrapped individually or in multiples using paper, nylon or foil autoclave bags.

5.6 INSECT CELL CULTURE

Insect cells grow optimally at 27–28°C, but lower temperatures (20°C) are useful for keeping the cells as a slower growing stock. Sf cells appear to grow at any temperature between about 18°C and 30°C! Above about 35°C though, a heat-shock response is invoked and it is not recommended that cells are incubated above 28°C. Under optimal conditions, population doubling times are usually 20–24 h at 27–28°C, in media supplemented with 10% FCS. For most work with the baculovirus expression system, the insect cells will be maintained and used as monolayer cultures in either plastic or glass flasks and plastic Petri dishes. The cells do not require CO_2 incubators and any standard laboratory incubator capable of maintaining 20–28°C is suitable; cells can also be incubated on the bench at room temperature or in constant temperature rooms.

On occasion larger spinner cultures of cells are useful and these can be as simple or elaborate as finances allow. Most economically, the cells can be cultured in rounded, flat-bottomed pyrex-flasks in volumes from 40–500 ml. A simple magnetic-stirring bar, slow-speed stirrer and pleated-foil cover are all that is required, as shown in Figure 5.1a. Alternatively, several vessels are available commercially that allow the culture of cells up to 1 l, without the need for air-sparging. One that we have tested and use extensively is the Techne-biological culture flask system, consisting of three different sized flasks and a special stirring base, allowing culture volumes from about 40 to 1000 ml (see Figure 5.1b). As an alternative to spinner cultures, cells may be maintained in suspension using shake flasks, and an orbital shaker normally used for bacterial cultures. Sterile, disposable shake flasks can be obtained from Bibby.

A laminar flow cabinet is the minimum requirement for routine sub-culturing of cell lines. Although if routine cell culture and virus infections are undertaken in the same laboratory it is preferable to use a Class II microbiological safety cabinet for cell culture work, to prevent virus contamination of stock cell lines. Most virus work can safely be carried out on the bench using 'good microbiological practice', although a Class I or II safety cabinet is useful when harvesting virus-infected spinner cultures. The exception to this rule is if the protein that is expressed from a recombinant virus is itself harmful or infectious. In these cases the appropriate safety guidelines must be adhered to (see the appropriate ACDP and ACGM or NIH guidelines). It is not advisable to undertake routine cell culture and virus work in the same cabinet, in fact it can be recipe for disaster! Equally, it is advisable that you should get into a daily routine of performing all cell culture work (e.g. passaging stock cells) before undertaking any work with virus or virus-infected cells. This minimizes the problems of accidently contaminating stock cells with virus.

For long-term storage of cells, liquid nitrogen facilities are required. Healthy low-passage number cells should be transferred to liquid

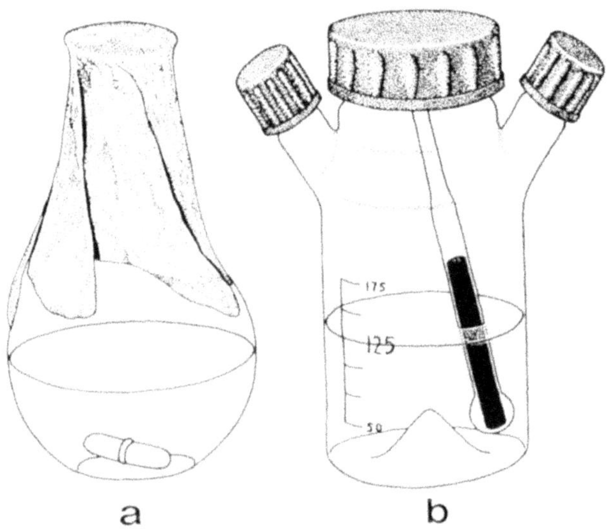

Figure 5.1 Schematic illustration of (a) a 'home-made' spinner culture flask and (b) a commercially available spinner flask (Techne biological stirrer).

nitrogen storage containers as soon as possible after receiving cell lines. To our knowledge insect cells survive indefinitely at these temperatures, certainly for many years.

In this section the use of TC100/FCS growth medium is recommended in all protocols. This medium can be substituted by any of the media described above. Some general points about insect cell culture follow and should be read before starting work.

* Fresh culture medium should, preferably, be warmed to room temperature before use.
* When removing medium from flasks or dishes, hold the vessel on one side so that the medium drains away from the cell monolayer. Remove the medium with a pipette, avoiding the the cell monolayer.
* When adding medium to cell monolayers in flasks or dishes, hold the vessel at an angle so that the medium is not pipetted directly on to the cells.
* When washing cell monolayers take great care not to dislodge the cells.
* When rocking dishes or flasks to obtain an even coverage of cells, do so gently in a backwards and forwards, left to right fashion (three or four times); do not use a swirling movement.
* In order to obtain healthy, viable cells when passaging stock cells or setting up for experiments, harvest the cells very gently. Three different methods seem to be favoured; firstly, you can tap the flask to dislodge the cells; secondly, you can pipette media rapidly over the surface of the cells; or finally, you can use a home-made cell scraper very gently (commercial ones are too harsh). Do not use trypsin or other enzymes to dislodge the cells.

5.6.1 ROUTINE SUB-CULTURING OF Sf CELL LINES (MONOLAYER CULTURES)

Required:

TC100 growth medium (see section 5.4.1)
Sterile cell scraper (optional; see below)
Plastic flasks (e.g. Gibco, Costar, Bibby, Sterilin)
and/or sterile glass flasks (e.g. ICN-Flow)
Container of hypochlorite or other disinfectant for receiving used media
Counting chamber (e.g. improved Neubauer)
Inverted microscope with ×10 and ×40 objectives
Benchtop centrifuge (optional)
Sterile universal bottles (optional)

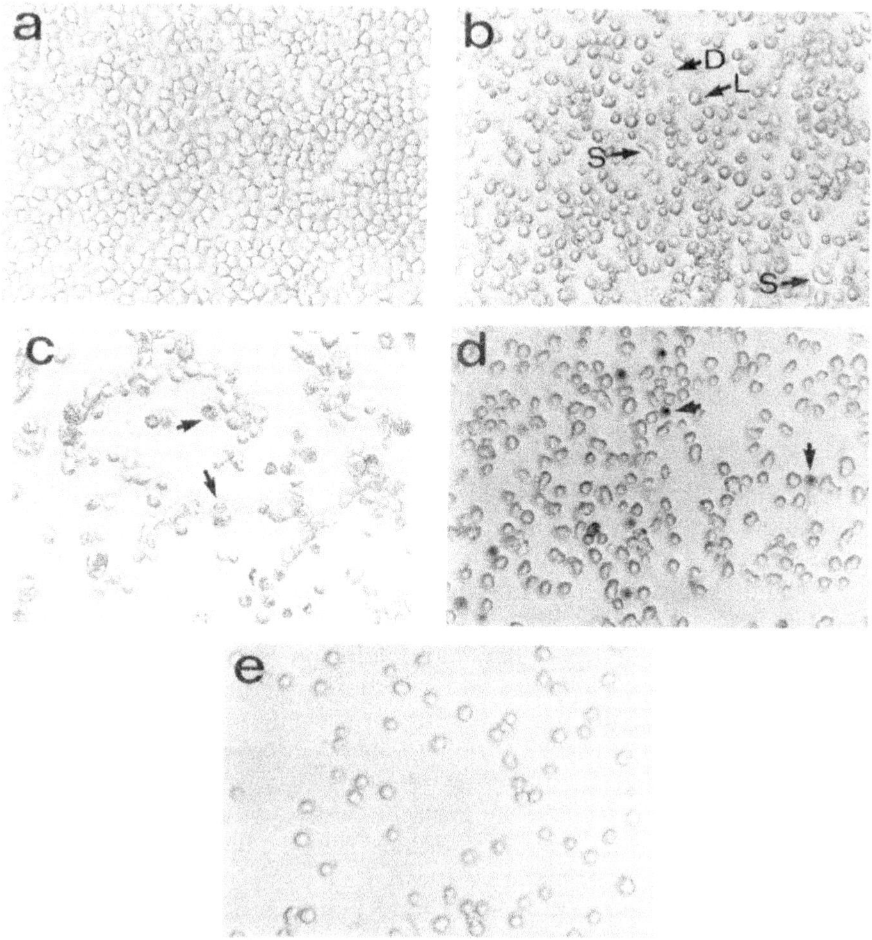

Figure 5.2 Plate (a), Sf cells that have formed a confluent monolayer and are ready for passaging. Plate (b), Sf cells that have been allowed to become over-confluent. Characteristic 'sausage' shaped cells (S) are indicated by arrows. These cells were photographed under phase contrast optics and clearly show the difference between live (L) and dead (D) cells. Plate (c), Sf cells that have been accidently contaminated with AcMNPV. Polyhedra within the cell nuclei are indicated by arrows. Plate (d), assessing viability using trypan blue staining. Live cells exclude the stain and dead cells (arrowed) take up the stain. Plate (e), Sf cells showing typical cell density for routine passaging, at weekly intervals.

10 ml and 1 ml pipettes (disposable or sterile glass)
0.2% (w/v) Trypan blue (optional; Sigma)
* Work in a laminar flow cabinet, or Class II safety cabinet. Swab all work surfaces and the outside of media bottles with 70% alcohol before starting work. Wash hands before and after working.

Method:

1. Inspect cell monolayers under the inverted microscope. Cells ready for sub-culturing will have formed a just-confluent monolayer of rounded cells (see Figure 5.2a).
   * If cells are overdue for passaging, many will be floating in the medium (see Figure 5.2b); you can still sub-culture at this stage but do not let this happen too often. Do not use such cells for experimental work. You will also notice the appearance of cells with a characteristic 'sausage' shape, caused by the cells dying during cell division (see Figure 5.2b)
   * Sf cells adhere more firmly to plastic surfaces than to glass.
   * It is good practice to look at cells under higher magnification (×40) during passaging to check for contamination by micro-organisms such as bacteria or yeast. Signs of accidental AcMNPV-infection include a failure of cells to form a confluent monolayer, failure of cells to attach well to either glass or plastic surfaces and the presence of polyhedra in cell nuclei (see Figure 5.2c). Improperly cleaned or rinsed glassware will also result in cells failing to attach and grow properly (see section 5.5).
2. If the cells are firmly attached to the flask, decant the used medium and pipette a small volume (about 5 ml) of fresh growth medium into the flask. If the cells are floating in the medium then omit this step. In the latter situation, the cells can be concentrated by pelleting at 500–800 r.p.m. for 5 min in a bench-top centrifuge and resuspending in fresh growth medium, although this is not normally essential for routine sub-culturing.
3. If the cells are firmly attached, dislodge them into the fresh medium by either tapping the flask sharply on the bench or against your hand, by rapidly pipetting medium over the cells, or by gently scraping the cells with a sterile cell scraper. Do not use trypsin. It is very important that you then separate any clumps of cells by gently pipetting up and down several times, taking care not to introduce air-bubbles.
   * Cell scrapers are easily prepared by cutting sections of glass rod (4–5 mm diameter) about 30 cm in length and bending the

final 1 cm to an angle of about 45°, using a Bunsen flame. Attach about 2 cm soft silicone tubing to the bent end. Place individual or several scrapers together in a paper or nylon autoclave bag (or wrap in foil) and sterilize by autoclaving. Dry before use. After use scrapers should be soaked in detergent and rinsed thoroughly under running water and then in distilled water (see section 5.5). Cells infected with virus do not need scraping to detach, but if scrapers do come into contact with virus-infected cells, they should be soaked in a suitable disinfectant (e.g. hypochlorite) before cleaning.

4. Remove a sample of cells for counting and estimating viability. Cells may be counted using a standard counting chamber.
   * To use the counting chamber, clean the chamber and cover-slip with alcohol, dry and fix cover-slip in position. After harvesting and de-clumping the cells, remove a sample and apply to the counting chamber; do not overfill. Observe cells under the ×10 objective on the inverted microscope, using phase-contrast if possible (see note below). The central large gridded square represents 1 mm^2 (see Figure 5.3), so by counting the number of cells in this area (which is usually divided into 25 smaller squares) and multiplying by 10^4 gives an estimate of the number of cells per ml. Always prepare duplicate samples and average the count.
   * Viability should be assessed when counting cells, either by using phase-contrast optics or by using the trypan blue exclusion test. Under phase-contrast, live cells appear bright and have a 'halo' around them, whereas dead cells appear darker and have a grainy appearance, as shown in Figure 5.2b. Alternatively, before applying to the counting chamber stain a sample of cells with an equal volume of 0.2 % (w/v) trypan blue. Live cells exclude trypan blue whereas dead cells take up the stain and appear dark blue, as shown in Figure 5.2d. Remember to multiply your cell count by 2 if staining with trypan blue.
   * If the cells are too numerous to count directly, dilute the cell suspension 1 in 5 or 1 in 10 with medium, before counting; multiply by the appropriate dilution factor afterwards.
   * After careful harvesting, healthy monolayer cells from glass flasks should give a viability of at least 90% whereas from plastic flasks it is about 80–90%. From spinner cultures (see next section) viabilities are usually higher (above 95%).
5. Add growth medium to fresh flasks using the following guide to approximate volumes. Swirl the media to wet the surface of the flask that the cells will attach to.

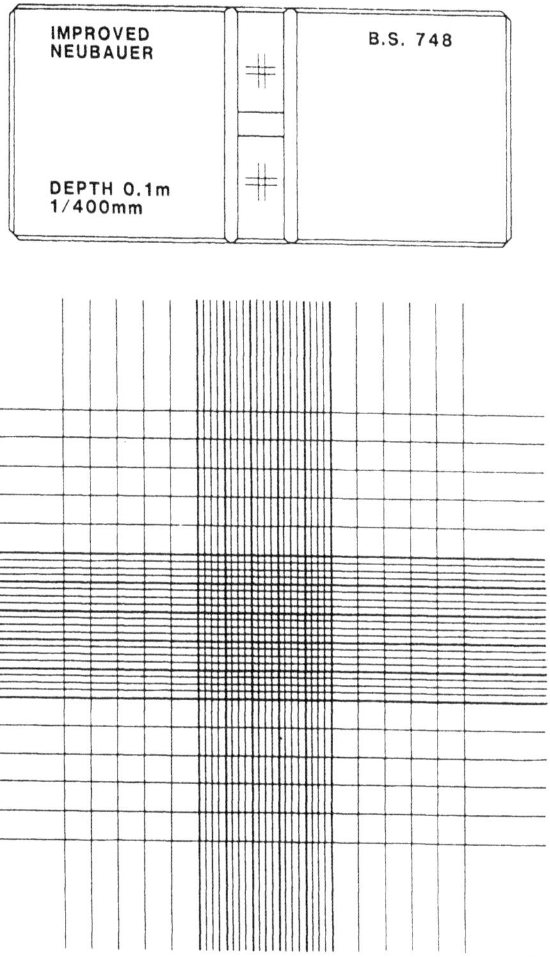

Figure 5.3 Diagrammatic representation of the counting area of an improved Neubauer counting chamber. The number of cells in the central large gridded area (1 mm^2 × 0.1 mm depth; divided into 25 medium-sized squares) are counted. The count is multiplied by 10^4 to obtain the number of cells/ml.

25 cm^2 flask (plastic)	4–5 ml
75/80 cm^2 flask (plastic)	10 ml
Small (75cm^2) glass flask (ICN-Flow)	15–20 ml
Large (120 cm^2) glass flask (ICN-Flow)	30–50 ml

6. After counting, add cells to the prepared flasks using the following guidelines:

25 cm^2 flask	$0.5–1.0 \times 10^6$

75/80 cm^2 flask	1–2 × 10^6
Small glass flask	1–2 × 10^6
Large glass flask	3–4 × 10^6

7. Cap flasks tightly and lay on their sides. Rock the flask gently to ensure an even coverage of cells. Label flasks with date, your initials, cell line and passage number. The number of cells used to seed the flask can also be recorded.
8. With a little experience it becomes evident that each week approximately one tenth of the cells in a flask need to be seeded into the next flask to continue sub-culturing at weekly intervals. It is therefore possible to omit counting the cells and simply add 1/10 of the cells harvested into the next flask (1/10 split), however, this assumes that all is well and that the cells have reached confluence. You should always check the cells under the microscope at each passage, even if you do not count them.
   * Depending on the temperature of incubation (20–28°C), the cells will be ready to be passage again within 5–7 days. It is best to adjust either cell numbers or temperature so that you passage your stock cells once per week. If extra cells are required for experimental work at an earlier date, put up additional flasks seeded at a higher density. Figure 5.2e shows cells that have been seeded at a typical density for weekly passaging.

5.6.2 ROUTINE SUB-CULTURING OF Sf CELLS MAINTAINED IN SPINNER CULTURES

As an alternative to sub-culturing Sf cells as monolayer cultures, spinner (or shaker) cultures can be used. In many ways these are easier and yield more reproducible cells each week than monolayer cultures. Spinner cultures are also highly recommended for the production of large stocks of virus for inoculum and for the subsequent purification of virus particles and DNA (see Chapter 6). Always adapt cells growing in monolayer culture to spinner culture conditions for a week or two before preparing large-scale cultures for virus infection.

To establish and maintain small-scale spinner cultures.

Required:

Sterile, rounded, flat-bottomed flasks with magnetic stirring bar and pleated foil cap (capacity 250 ml), or,
Sterile specialist-type flask (e.g. Techne 125 ml biological stirring flask), or,
Sterile shake flasks (250 ml)
Sf cells harvested from a monolayer culture (see section 5.6.1)

TC100 growth medium (see section 5.4.1)
Magnetic stirrer capable of low speeds (50–75 r.p.m.) or,
Shaking platform/orbital shaker for shake flasks
Incubator capable of taking magnetic stirrer, etc., warm room or water bath-jacket, at 25–28°C.

Method:

1. Using aseptic technique, transfer TC100 growth medium and Sf cells into a sterile flask to give a final volume of 50 ml at 5×10^5 cells/ml.
   * Glass flasks must be thoroughly cleaned and rinsed (see section 5.5). 'Home-made' spinner-flasks can be prepared from pyrex flat-bottomed, round-bodied flasks. Place a clean magnetic stirring bar inside before autoclaving and use a double layer of pleated tin-foil as a cover (see Figure 5.1a). Do not overfill flasks (i.e. not more than 100 ml in a 250 ml flask) as the culture medium should have the maximum surface area for gas exchange. Commercial specialist flasks can also be used, such as the Techne biological stirrer (see Figure 5.1b). Shake flasks can either be made of glass or disposable plastic flasks for tissue culture use can be purchased (e.g. from Bibby).
   * Seeding the initial culture at 5×10^5 cells/ml ensures that the cells establish in spinner culture quickly. Once established, subsequent spinner cultures are seeded at $1–2 \times 10^5$ cells/ml.
2. Place cells at 25–28°C and stir, or shake, relatively slowly (50–75 r.p.m.) until the cell density reaches $2–3 \times 10^6$ cells/ml. Sample cells daily for counting.
   * If clumping of cells is a problem, stir the cells at a higher speed, up to 100 r.p.m. In this case the use of pluronic F-68 as a surfactant to reduce shear stress may be beneficial (see section 5.4.6). Higher cell densities, up to 5×10^6 cells/ml, have been reported using EX-CELL 400 and pluronic (Maiorella *et al.*, 1988).
3. When the cells reach a density of $2–3 \times 10^6$ cells/ml, they can be used to seed a fresh spinner or shake culture, e.g. 50 ml at $1–2 \times 10^5$ cells/ml, adding fresh TC100 growth medium as appropriate. Alternatively, the same flask can be used by removing the excess cells and topping up to volume with fresh growth medium; in this way the same flask can be used for several weeks.
   * It is useful to plot a growth curve by sampling the cells daily for total and viable counts; a typical example is shown in Figure 5.4. The cells will continue to divide exponentially until they reach about $2–3 \times 10^6$ cells/ml, after which time the proportion of dead cells increases.

* It is recommended that the spinner or shake culture be 'passaged', i.e. taken down to 1–2 × 10^5 cells/ml, when the cells reach a density of 2–3 × 10^6 /ml. This is usually required every 6–7 days at 25°C and every 5–6 days at 28°C (when using FCS at 10%). If you are using other media, we recommend that you prepare your own growth curve to assess the best time for harvesting cells for experimental work or the next passage.

4. Cells required for experimental work, for example plaque-assays and virus infections in monolayer cultures, are simply removed from the spinner or shake culture and seeded at the appropriate density (see section 5.7) in Petri dishes or flasks. After 1–2 h, the cells should have attached to the Petri dish or flask and will be ready to infect.

Medium-scale spinner cultures (100–500 ml).

Required:

1 l flat-bottomed, round-walled flask (prepared as described above),
or,
1 l specialist-type flask (e.g. Techne)
TC100 growth medium
Sf cells harvested from a spinner or shake culture (not from monolayer cultures), as described above

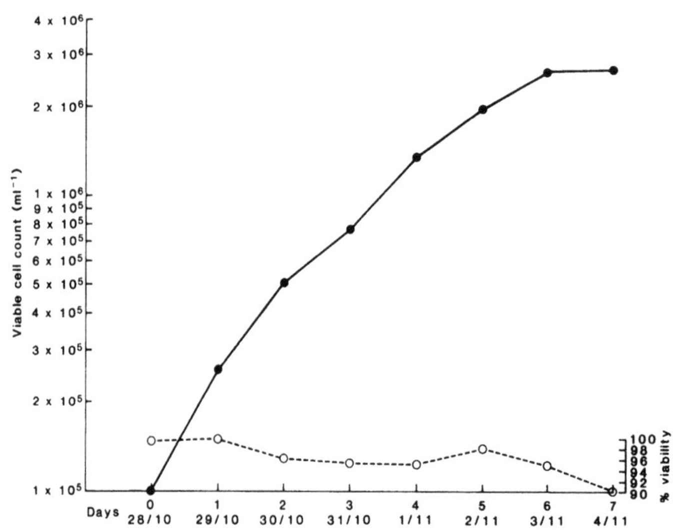

Figure 5.4 A typical growth curve prepared using Sf21 cells growing in TC100/5% FCS in a 100 ml Techne biological stirrer.

Table 5.4 A guide to *S. frugiperda* cell seeding densities for experimental work

Dish/flask size	*Seeding density*[1]	*Volume of medium*	*Uses*
30–35 mm dish	*1×10^6 (1.5×10^6)*	*1.5–2.0 ml*	*Plaque-assays* *Virus infections for protein/DNA analyses*
50–60 mm dish	*2×10^6 (2.5×10^6)*	*3.0–4.0 ml*	*Plaque-assays* *Virus infections for DNA Analysis* *Amplification of small stocks of virus*
25 cm^2 flasks	*1×10^6 (1.5×10^6)*	*3.0–4.0 ml*	*Amplification of small stocks of virus*
75 cm^2 flasks	*$0.5–1.0 \times 10^7$*	*10 ml*	*Amplification of medium-scale stocks of virus* *Virus infections for RNA analysis*
Spinner cultures	*$1–2 \times 10^5$/ml*[2]	*40–500 ml*	*Amplification of high-titre stocks of virus* *Medium-scale production of virus for virion and DNA purification* *Medium-scale production of recombinant protein*

[1] *Figures in brackets refer to seeding densities to be used with a 2 h incubation period prior to virus infection, otherwise leave the cells overnight before use.*
[2] *Grow cells to a density of 5×10^5/ml prior to infection for high-titre virus stocks and $1–2 \times 10^6$/ml for the production of recombinant protein.*

Method:

1. Medium-scale spinner cultures can be prepared from a stock 50 ml spinner or shake culture by simply seeding cells at $1–2 \times 10^5$/ml in the desired volume (up to 500 ml in a 1 l flask), and allowing the cells to reach the appropriate density before, for example, infecting with virus. To achieve high-titre virus stocks for experimental use (as inoculum, for example) the cells should be infected when they reach a density of 5×10^5 cells/ml. This will be discussed more fully in the next chapter.
   * For large-scale cultures above 500 ml it is recommended that you use special flasks that allow oxygen sparging (see Chapter 10). Anti-foam agents and surfactants (e.g. pluronic F-68) can also reduce the number of battered cells, as insect cells are sensitive to

shear forces in spinner culture (Tramper *et al.*, 1986; Murhammer and Goochee, 1988).

2. Addition of pluronic F-68 to 0.05–0.01% (determine best concentration empirically) may increase the percentage of viable cells and overall cell density attainable.

5.7 A GUIDE TO Sf CELL SEEDING DENSITIES FOR EXPERIMENTAL WORK

Table 5.4 gives suggested initial seeding densities for Sf cells for a variety of experimental procedures. Cells can either be harvested from suspension cultures or from monolayer cultures. Where indicated, 'overnight' implies at least 12 but not more than 24 h incubation at 28°C.

5.8 FREEZING, STORAGE AND RECOVERY OF INSECT CELLS IN LIQUID NITROGEN

This section describes protocols for the successful freezing and thawing of insect cells and their storage in liquid nitrogen. On receipt of insect cells from other laboratories or Culture Collections, you should aim to freeze low-passage number stocks as soon as possible. A stock of 20–30 vials should be adequate. After 1–2 weeks of storage, it is worthwhile retrieving one vial of cells to check on their viability.

5.8.1 FREEZING AND STORAGE OF CELLS IN LIQUID NITROGEN

Required:

Healthy, just-confluent Sf cells (at least 90–95% viable)
TC100 with 10% FCS and antibiotics
Tissue-culture grade, filter-sterilized, dimethylsulphoxide (DMSO, e.g. Sigma)
Cryogenic vials (for 1 ml samples)
Freezer canes/sleeves, if required (see below)
Liquid nitrogen storage facilities
−20°C and −70–80°C freezers, if required (see below)

Method:

1. Harvest healthy Sf cells, from suspension or monolayer cultures, into a small volume of fresh growth medium containing 10% FCS. This may be achieved by pelleting the cells at a slow speed

(500–800 r.p.m.) for 5 min and resuspending in fresh growth medium (10% FCS).

* If freezing cells adapted to growth in serum-free medium, cell-culture grade bovine serum albumin (e.g. from Sigma) should be added to a final concentration of 1%.

2. Determine the cell density and viability. Only cells that are at least 90–95% viable should be used for freezing.
3. Adjust the cell concentration to $2–3 \times 10^6$ cells/ml with growth medium/10% FCS.
4. Prepare a freezing mix consisting of: 80% (v/v) growth medium/10% FCS and 20% (v/v) DMSO. It is important to add the DMSO to the media (not the other way around) and mix in quickly. Chill to 4°C.
5. Prepare 1 ml cryogenic vials by labelling with date, cell type and passage number. If required, have freezer canes and cardboard/plastic sleeves ready.
6. Add an equal volume of freezing mix to the Sf cell suspension and mix (final DMSO concentration = 10%). Place 1 ml aliquots into the prepared vials. It is important that the cells are now chilled immediately as DMSO is cytotoxic.
7. If your liquid nitrogen container has the facility for freezing slowly in the vapour phase, use this to chill the cells overnight and then transfer to the liquid phase (using canes and outer sleeves if required).

   If you have no vapour-phase chilling facility, the following procedure works equally well. After adding the freezing mix to the cells, place the vials in the racks or canes that they will be stored in. Place tubes, upright, in a −20°C freezer for 1–2 h and then transfer, in an ice-box, to a −70–80°C freezer for 4–6 h, or overnight. Transfer promptly to the liquid nitrogen storage facility.

   * Some protocols for freezing insect cells recommend purchasing an automatic chilling device that slowly lowers the cells from the vapour into the liquid phase of the nitrogen storage tank. These are very expensive and from our experience unnecessary.

   * We recommend that you retrieve a vial after one to two weeks, to determine the success of your freezing/thawing regime.

   * Keep records of all frozen stocks of cells.

   * Always wear protective cryogenic gloves and face mask when placing cells into liquid nitrogen.

5.8.2 RECOVERY OF CELLS FROM LIQUID NITROGEN

Required:
Vial of cells stored in liquid nitrogen
TC100 medium with 10% FCS
25 cm^2 flask
Water bath at 37°C (optional)
70% alcohol

Method:
1. Remove a vial of cells from the liquid nitrogen storage facility. Wear protective cryogenic gloves and a face mask.
   * Keep a record of which cells have been retrieved.
2. Thaw cells rapidly by warming in a gloved hand or by holding the vial in a water bath at 37°C. To avoid contamination, ensure that only the part of the vial containing the cells is in the water.
   * Some workers prefer to immerse the vial in a container of 70% alcohol, standing in a water bath at 37°C.
3. Immerse, or thoroughly swab, vial in 70% alcohol and open carefully in a laminar flow hood.
4. Pipette the cell suspension into a 25 cm^2 flask containing 5 ml growth medium (10% FCS). It may help if the medium has been pre-warmed to 28°C.
5. Incubate the flask at 28°C until the cells have recovered and attached to the surface (1–2 h is usually sufficient time). Because of the presence of the DMSO, do not leave cells more than 24 h at this stage. Remove the medium (which will contain DMSO and any dead cells) and replace with fresh growth medium (10% FCS). Incubate the cells at 28°C until they have formed a confluent monolayer.
6. Sub-culture the cells into two 75 cm^2 flasks and continue routine sub-culturing as described in section 5.6.

5.9 A GUIDE TO ADAPTING CELLS TO SERUM-FREE MEDIA

If you routinely culture Sf cells in medium supplemented with FCS and wish to change to serum-free medium, the cells will usually need to be weaned onto the new media over a period of time. We suggest the following protocol, that has worked in the adaptation of several insect cell lines including Sf21 and Sf9 cells to Sf900, EX-CELL 400 and IPL-41 media.

Required:
Healthy cells (low passage number) growing in medium supplemented with serum

Serum-free medium (see comments in 5.4.6)
Disposables for cell culture work

Method:

1. At the next routine passage, transfer cells into medium consisting of 75% serum-containing and 25% serum-free. Allow the cells to become confluent. This may take longer than usual.
2. At the next passage, transfer the cells into medium consisting of 50/50 with and without serum. Allow the cells to grow and become confluent. If the cells are very slow to grow, at the next passage repeat using the medium as a 50/50 mixture.
3. At the next passage transfer the cells into medium consisting of 75% serum-free and 25% containing-serum. Allow the cells to grow until confluent.
4. Repeat step 3 if the cells are still growing quite slowly.
5. Transfer the cells into 100% serum-free medium. Watch carefully, they will probably take two or three passages to grow to optimum densities.
6. As soon as possible, freeze samples of cells adapted to the serum-free medium in liquid nitrogen.

Propagation, titration and purification of AcMNPV in cell culture 6

6.1 INTRODUCTION

Success with the baculovirus expression system is dependent on being able to infect cells efficiently with AcMNPV, thus obtaining maximum virus replication and hence optimum production of the desired protein. Before being able to create a recombinant virus in the first place you must have stocks of infectious, purified virus DNA into which you will insert the foreign gene. You will need to obtain the appropriate virus stock (AcMNPV or AcMNPV.*lac*Z) to amplify the virus and to purify virus particles from which the DNA can be isolated. All of these procedures are quite straightforward and relatively easy to undertake in the laboratory, if a few simple guidelines are adhered to.

The titre of all virus stocks (wild-type or recombinant) should always be determined accurately by plaque-assay (Brown and Faulkner, 1977). A less satisfactory method for determining the virus titre is by tissue culture infection dose ($TCID_{50}$), the method for which is given at the end of this chapter. Plaque-assays will also need to be carried out to separate recombinant and wild-type virus after a co-transfection. Thus being able to perform an accurate and successful plaque-assay is very important. We strongly recommend that you carry out one or two practice plaque-assays using AcMNPV (or AcMNPV.*lac*Z) before using this technique to screen for recombinant virus. Having isolated plaques for putative recombinant viruses, you need to amplify the viruses and test them for expression. These stages require that you know how to infect cells with virus correctly and know, for example, when to use low multiplicities of infection (moi) and when to use a high moi.

This chapter describes all the basic techniques that you will need for infecting cells with virus, virus amplification and purification, and includes protocols for:

- virus infection of cells in monolayer culture, for experimental work
- virus infection of cells in suspension culture, for experimental work and the production of virus stocks
- titration of virus by plaque-assay
- plaque-purification of virus
- amplification of virus from single-plaque picks through to medium-scale suspension culture
- amplification of virus stocks obtained as gifts or purchased
- purification of virus particles
- virus DNA purification
- titration of virus by $TCID_{50}$

6.1.1 SAFETY CONSIDERATIONS: GENERAL RULES FOR WORKING WITH BACULOVIRUSES

Some general rules for working with baculoviruses follow and should be read before proceeding with the experimental work:

1. Take care to keep stock cell lines and virus-infected cells separate. Never work with virus in the same hood that is used to maintain cell stocks.
2. Try to carry out all routine passaging of cells early in the morning, before handling virus-infected materials (see Chapter 5).
3. It is quite safe to work with virus-infected cells on the bench using good aseptic and microbiological techniques. However, it is wise to perform all manipulations that may cause large amounts of aerosols (e.g. harvesting suspension cultures of virus-infected cells) in a Class I or II microbiological safety cabinet.
4. It is always good practice to work over benchkote, have a disinfectant available in case of spills and to wear gloves at all times. A discard pot containing disinfectant, e.g. hypochlorite, should always be to hand for virus-contaminated liquid waste. All disposable waste materials should be placed in Bio-hazard bags for autoclaving. Never pour virus-containing solutions into the drains without prior disinfection. Contaminated glassware should be autoclaved.
5. When performing any work with virus-infected cells, remember to treat culture media and washings from cell pellets, etc. as contaminated with virus (ECV) and dispose of safely. Cell-released virus particles may be inactivated by 70% ethanol, chemical disinfectants or by autoclaving. Polyhedra and cell debris containing polyhedra are best destroyed by autoclaving or by the use of hypochlorite.

6. Hands should be washed before starting work and must be washed thoroughly afterwards. Benches should be disinfected after each work period with, for example, 70% ethanol or Virkon (1% solution). If Gilson pipettes are used to dispense virus-infected samples, they should be regularly decontaminated by soaking the barrel in detergent or disinfectant solution. Gilson pipettes can be a prime source of possible cross-contamination problems.

6.2 INFECTION OF CELLS WITH VIRUS FOR EXPERIMENTAL WORK

6.2.1 INFECTION OF Sf CELLS IN MONOLAYER CULTURE

This section gives general guidance notes on how to infect monolayers of Sf cells, for example, to analyse virus-induced proteins, RNA, or DNA. More detailed protocols for the extraction and analysis of these molecules are given in Chapter 8. Most work can be conveniently carried out by infecting cell monolayers prepared in 35 mm Petri dishes (tissue culture grade), however, if increased amounts of material for analysis are required, you can infect cells prepared in 60 mm dishes or 25/75 cm^2 flasks. The main protocol below is written for 35 mm dishes and a table is provided at the end of the section (Table 6.1) to give equivalent volumes for working with other sizes of dishes and flasks.

Required:
- 35 mm Petri dishes (tissue culture grade)
- Healthy cells (Sf21 or Sf9; see Chapter 5)
- TC100 growth medium or alternative (see Chapter 5)
- Virus inocula
- Sandwich box lined with moist tissue paper
- Incubator at 28°C
- Discard facilities for waste media

Table 6.1 Volumes to use when working with various culture dishes/flasks

Size of dish/flask	*Number of cells (2 h : o/n)*	*Inoculum volume (min : max)*	*Medium volume*
35 mm dish	1.5 : 1.0 × 10^6	0.1 : 0.5 ml	1.5–2.0 ml
60 mm dish	2.5 : 2.0 × 10^6	0.4 : 1.0 ml	3.0–5.0 ml
25 cm^2 flask	2.0 : 1.5 × 10^6	0.4 : 1.0 ml	3.0–5.0 ml
75 cm^2 flask	6.0 : 4.0 × 10^6	1.0 : 2.0 ml	10–15 ml

Method:

1. Seed 35 mm dishes with 1 × 10^6 cells in 1.5–2 ml growth medium and incubate overnight at 28°C (or seed with 1.5 x 10^6 cells and leave for 1–2 h, see table 6.1). Prepare dishes for each virus sample to be analysed and include two dishes for the controls; mock- and AcMNPV-infected.
   * Always include mock-infected and wild-type virus-infected cells as controls when analysing recombinant virus stocks; in this way you can check that your infection procedures are effective, as well as having useful controls for DNA, protein gels, etc.
   * Always check the cells under the microscope before infecting with virus, just to confirm that all is well, i.e. that cells have attached well and have formed an even monolayer that is not too sparse, overcrowded or clumped. If cells are too thin you will have insufficient material for analysis and if they are too crowded the virus will not replicate efficiently (and the moi will be wrong in both cases). If the cells have not settled down and attached to the Petri dish within 1–2 h, they are unlikely to do so, and you will need to prepare fresh cells.
2. Remove the medium from the cells and add virus inoculum at a multiplicity of 10 pfu/cell (moi = 10). If the volume required is less than 100 μl, add medium to increase the volume to this amount. If the volume required is over 0.5 ml the virus inoculum is really too dilute. Add inoculum dropwise, very gently to the centre of the dish.
   * To ensure synchronous infection of every cell, addition of virus to 10 pfu/cell is recommended, however, 5 pfu/cell (moi = 5) should also work.
3. Incubate the cells for 1 h at room temperature; ensure the surface is level. If preferred, you can gently rock the dishes every 15–20 min to ensure an even coverage of virus over the cells. In practice this does not appear to be necessary when infecting cells in 35 mm dishes; it is necessary when infecting cells in 25 or 75 cm^2 flasks.
4. After the 1 h incubation, remove the inoculum and discard safely. Tilt the dish to one side and use a Pasteur pipette to aspirate the medium. Replace immediately with 1.5–2.0 ml TC100 growth medium. Again pipette gently so as not to dislodge the cells.
5. Incubate the cells in a humidified sandwich box (line with moist tissue paper) at 28°C for the required time period. See detailed protocols in Chapter 8.
6. Refer to Table 6.1 when working with other sizes of flasks/dishes.

6.2.2 INFECTION OF Sf CELLS IN SUSPENSION CULTURE

This section gives general guidance on the infection of cells that are growing in suspension culture. The technique is much the same whether you are infecting the cells for experimental use (e.g. to purify recombinant proteins) or for the amplification of virus stocks. The difference comes in the moi used; in the former case a high moi is normally required and in the latter a low moi must be used.

Required:
- Flask of Sf cells in suspension culture
- TC100 growth medium or alternative (see Chapter 5)
- Virus inoculum
- Discard facilities

Method:
1. Seed a suspension culture flask with cells at 1–2 × 10^5/ml in TC100 growth medium; volume as required (usually 50–500 ml). Grow the cells until they reach the required density: 5 × 10^5/ml for preparing virus stocks; 1–2 × 10^6/ml for the analysis of recombinant proteins.
   * See Chapter 5 for more details on growing cells in suspension culture.
2. Calculate the required amount of virus inoculum. For the preparation of stocks of virus inocula, infect cells at a low moi (0.1–0.2 pfu/cell) and for the analysis of recombinant proteins, infect cells at a high moi (5–10 pfu/cell). Simply add the required amount of virus inocula to the flask of cells and return to the stirrer.
   * Although it is sometimes recommended, we have found no advantages in concentrating the cells by centrifugation prior to infection, and then diluting again afterwards.
3. Incubate the cells at 28°C for the required time. Harvest the medium for use as virus inoculum after 4–5 days, or when the cells are well infected. Alternatively, harvest cells/media for the analysis and/or purification of recombinant protein, as required.
4. Medium harvested as a source of virus for inoculum should be dispensed into small aliquots and stored at 4°C, in the short term. AcMNPV inocula store well at 4°C for about 1 year, after which time the virus titre drops steadily. For long-term storage, virus inocula should be kept at −70–80°C, in small aliquots in glass or cryogenic vials.
5. Before using the virus inocula in experimental work, you should determine the titre by plaque-assay (see section 6.3)

6.3 TITRATION OF VIRUS BY PLAQUE-ASSAY

Plaque-assays are carried out to determine virus titres in plaque-forming units per ml (pfu/ml) so that standard amounts of virus can be used to infect cells during experimental work. Plaque-assays are also used to titrate and separate recombinant and wild-type virus following co-transfections (see Chapter 7). The basic protocol for AcMNPV is given first, followed, a modified assay for *lac*Z-positive viruses, where plaques are stained with X-gal to give a blue-coloration. It is worth mentioning that, in our experience, Sf21 cells give more well-defined plaques than Sf9 cells.

6.3.1 STANDARD PLAQUE-ASSAY

Required:

35 mm Petri dishes seeded with Sf cells (see point 1)
TC100 growth medium (with 10% FCS and antibiotics)
2% (w/v) low-gelling temperature agarose in water (Sea-plaque, from FMC or Sigma type VII); sterilized by autoclaving
Virus inoculum to be titrated
0.5% (w/v) neutral red in water (Sigma); filter-sterilized and stored in the dark
Phosphate-buffered saline (PBS, prepared from tablets or powder and sterilized by autoclaving; see Table 5.1)
Disposal facilities for virus-contaminated media/materials
Sandwich box with moist tissue paper and an incubator at 28°C

Method:

1. Prepare 35 mm dishes of cells seeded at 1.5×10^6 in 1.5–2.0 ml medium and leave 1–2 h (or at 1×10^6 and leave overnight) at 28°C, prior to starting the assay. Ensure cells have settled down to form a fairly sparse, even monolayer (50% confluence). If the cells become confluent too soon, very tiny plaques (or even none at all) will form.
2. Working on the bench near to a bunsen, or in a Class I/II microbiological safety cabinet, prepare serial log (1 in 10) dilutions of the virus to be titred, using growth medium as the diluent. You will require 100 μl of each dilution for each dish used. Use sterile bijoux or microcentrifuge tubes etc. as appropriate.
   * The dilutions to prepare will depend on the source of virus to be titred; wild-type AcMNPV will replicate to a titre of about 1–2 x 10^8 pfu/ml (plate 10^{-5} to 10^{-7} dilutions); recombinant viruses to about 10^7 to 10^8 (plate dilutions as for wild-type virus); and co-transfection supernatants will contain virus between

10^2 and 10^4 pfu/ml (plate neat to 10^{-4} dilutions). The aim is to obtain between 10–30 plaques on one of the dishes that can be easily counted. This is especially important when screening for recombinant viruses.

* Ideally dilutions should be assayed in duplicate dishes and when screening for recombinant viruses, sufficient dishes should be prepared to give at least 200 plaques for inspection (approx. 10 dishes) (see Chapter 7 for more details).

* If the virus dilutions are to be left for any length of time they should be stored on ice.

3. Using a Pasteur pipette (attached to a vacuum line is most convenient), remove the medium from the cells and discard into disinfectant. Take care not to disturb the cell monolayer.
4. Overlay the cells with 100 μl of the appropriate dilutions, by adding dropwise to the centre of the dish. Take care not to dislodge cells. Leave the dishes at room temperature for 1 h, on a level surface, for the virus to adsorb to the cells. Rock the dishes very gently, to ensure even coverage of the virus, at 15–20 min intervals (optional).
5. During the 1 h incubation prepare the agarose overlay.

   * For each dish, 1 ml of agarose solution and 1 ml TC100/10%FCS are required. If the autoclaved agarose solution has set, melt the required amount in a microwave oven or boiling water bath and cool to 37°C in a water bath or incubator (ensure water bath is clean, as often this is the source of contamination when plaque-assays become infected with bacteria or fungi). Just before use add an equal volume of TC100/10%FCS, mix in carefully without causing air-bubbles. If prepared too soon, store at 37°C to prevent setting; after adding TC100 you cannot re-melt the agarose.
6. Carefully remove the virus inoculum from the dishes and discard into disinfectant.

   * Failure to remove the inoculum prior to adding the solid overlay may result in poor plaque-formation and/or the solid overlay may fall out when the dishes are inverted after staining.
7. Gently pipetting down the side wall of the Petri dish, add 2 ml of the prepared agarose overlay to each dish. Allow the agarose to set at room temperature.
8. When the agarose overlay has set (10–15 min), add 1 ml TC100/10%FCS to each dish (liquid overlay) and place in a sandwich box lined with moist tissue paper.

   * We find it is essential to use a liquid overlay to obtain well-defined plaques. The reason for this is probably that the liquid overlay provides additional nutrients to supplement the

half-strength TC100 medium in the solid overlay. Unfortunately, 2× concentrated TC100 cannot be prepared for use in the solid overlay as the salts precipitate from solution.

9. Incubate the cells at 28°C for 3–4 days.
   * Small plaques, sufficiently developed to count, will be visible after three days, however when trying to screen for recombinant virus it is wise to leave the dishes an extra day or two, to obtain larger plaques with many polyhedra present in the wild-type plaques. It is then easier to spot the polyhedra-negative viruses.
10. Plaques are more easily visualized by staining the cell monolayer with neutral red. Dilute the stock stain (0.5% w/v) 1 in 20 with PBS before use and add 1 ml to the liquid overlay in each dish. Incubate the cells at 28°C for 2–4 h and then decant the 2 ml stain solution into disinfectant. Invert dishes and store at room temperature for 2 h to overnight, in the dark. This allows the plaques to clear and become visible, as shown in Figure 6.1.
    * Live cells take up the neutral red stain so that plaques appear as clear patches in a background of red cells (see Figure 6.1).
    * Neutral red stain becomes very granular on exposure to light

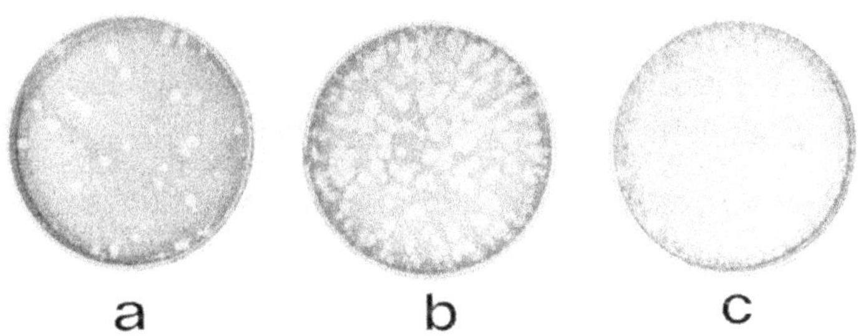

Figure 6.1 Titration of AcMNPV by plaque-assay. Monolayers of Sf cells are shown after staining a plaque-assay with neutral red (see section 6.2). Typical plaques are shown as clear areas in a stained (normally pink) background. The cells were infected with 100 μl virus inoculum at (a) 10^{-6} dilution, (b) 10^{-5} dilution and (c) 10^{-4} dilution. After counting the number of plaques in dish (a), the virus titre was estimated to be 3.4×10^8 pfu/ml (34 plaques counted $\times$ 10 $\times$ 10^6).

and this can present problems in trying to distinguish between polyhedra and grains of neutral red.

* To visualize *lac*Z-positive plaques please turn to section 6.3.2.

11. To determine the titre of a virus, examine the plates and find the dilution that gives a countable number of plaques (10–30 for example). Count the plaques from each dish at this dilution and obtain the average number, multiply by the dilution factor and by 10 (only 0.1 ml was applied to each dish) to give the titre in pfu/ml. Plaque numbers should follow the serial dilution.

    * For example, if there were an average of 15 plaques per dish for the 10^{-6} dilution, the virus titre would be $15 \times 10 \times 10^6 = 1.5 \times 10^8$ pfu/ml.
12. Hints for distinguishing recombinant plaques from wild-type are given in Chapter 7.
13. If plaques are to be picked for further rounds of plaque-purification or for virus amplification, do this as soon as possible using the protocols in section 6.4.
14. Discard all contaminated materials by autoclaving.

6.3.2 PLAQUE-ASSAY OF *lac*Z-POSITIVE VIRUSES

Required:

As for standard plaque-assay, plus
2% (w/v) X-gal in dimethylformamide (DMF) (store at −20°C)

Method:

1. Carry out a normal plaque-assay, as described above (section 6.3.1) until the staining step is reached.
2. To visualize *lac*Z-positive plaques, remove the 1 ml of liquid overlay from the dishes and discard into disinfectant. Replace with 1 ml TC100 growth medium containing 15 μl/ml 2% (w/v) X-gal.
3. Incubate the dishes at 28°C for at least 5 h (can be left overnight). *Lac*Z-positive plaques will turn blue. Drain the stain solution into disinfectant prior to viewing plaques.
4. If a mixture of *lac*Z-positive and *lac*Z-negative plaques are expected, the negative plaques can be more easily distinguished if the cells are dual stained with X-gal and neutral red. In this case, add 1 ml X-gal/TC100 as described above and incubate for 3 h at 28°C, then add 1 ml neutral red stain (prepared as described in section 6.3.1) and leave for an additional 2 h. Drain the stain solutions into disinfectant and allow plaques to clear by placing the inverted dishes at ambient temperature in the dark, for at

least 2 h before viewing. The blue coloration may take several hours to develop maximally.

5. *Lac*Z-positive or negative plaques are picked and re-titrated in the same way as wild-type virus plaques (see section 6.4).

6.4 PLAQUE-PICKING AND PLAQUE-PURIFICATION

After identifying plaques containing putative recombinant-virus, it is essential that several examples are picked and independently plaque-purified. The reason is that on occasion a false-negative may be picked e.g. a virus from which the polyhedrin gene is missing but which does not contain the foreign gene either. False-recombinants appear to be more common when using *lac*Z selection techniques (see Chapter 7).

Before picking plaques it is best, if you have not done so already, to ring the required plaques with a marker pen on the base of the Petri dish. Select plaques which are well isolated if possible. Plaque-picks can be used directly for the next round of plaque-purification; there is no requirement to amplify the virus in between rounds.

Required:
- Supply of sterile Pasteur pipettes and bijoux
- 35 mm dishes of Sf cells set up for a second plaque-assay
- TC100 growth medium
- Materials for a plaque-assay (see section 6.3.1)

Method:

1. Find well isolated plaques that you wish to pick and identify them on the base of the dish with a marker pen.
2. Take up a plug of agarose from the plaque into a Pasteur pipette and wash it out into 0.5 ml TC100 growth medium. Vortex to release the virus particles from the plug of agarose.
3. The 0.5 ml virus suspension can be stored at 4°C or it can be used immediately in the next round of plaque-assay or in virus amplification.
4. To re-plaque, prepare 10^{-1} to 10^{-3} dilutions (0.1 ml + 0.9 ml growth medium diluent) of the picked plaque suspension and use to infect cells for the next round of plaque-assay (see section 6.3.1/2). Prepare duplicate or triplicate dishes for each dilution. * The reason for preparing dilutions from 10^{-1} to 10^{-3} is that the yield of virus in any particular agarose plug may vary considerably.
5. For plaque-purification of recombinant or wild-type virus, at least three rounds of plaque-picking should be completed; in the case of purifying recombinant viruses from an initial mixture of

recombinant and wild-type, it is essential that plaque-purification is repeated until no wild-type plaques are found. If you are using the linear DNA method of co-transfection, it is often possible to obtain pure stocks of recombinant virus after only two rounds of plaque-purification (see Chapter 7 for more details).

6. After the final round of plaque-purification the virus needs to amplified. The method for this is described in section 6.5.

6.5 AMPLIFICATION OF VIRUS STOCKS

This section assumes that you are starting with a picked virus plaque in 0.5 ml TC100 medium. This is used to obtain a seed stock of virus from which intermediate and working stocks can be prepared.

Required:
Healthy Sf cells
TC100 growth medium or alternative (see Chapter 5)
Flasks for monolayer and/or spinner culture (see Chapter 5)

6.5.1 TO PREPARE A SEED STOCK OF VIRUS FROM A PLAQUE-PICK

Method:

1. Use about 200–250 μl of the virus suspension (0.5 ml), from a single plaque-pick, to infect 1–1.5 × 10^6 Sf cells in a 25 cm^2 flask or a 60 mm Petri dish. Remove the medium from the cells, add the inoculum and leave for 1 h at ambient temperature; gently rock the inoculum over the cell monolayer every 15–20 min.
2. Remove the inoculum and add 3–4 ml of fresh growth medium. Incubate the cells for 3–6 days at 28°C, until they are well infected with virus (check under the microscope for signs of infection).
   * Photographs showing non-infected Sf cells are shown in Figure 5.2, and cells infected with wild-type or recombinant virus are shown in Figure 6.2.
3. Harvest the medium and store at 4°C (see note below). This is your seed stock of virus. Store 1 ml of this seed stock at −70°C (no additives are required) and the remainder at 4°C.
   * The virus titre may be slightly increased by harvesting both the medium and cells for sonication (3 min in a water-bath type sonicator) prior to storage at 4°C.
4. Optional: titre your seed stock by plaque-assay (see section 6.3.1/2). The titre should be in the region of at least 10^7 pfu/ml.
   * If preparing a stock of recombinant virus, this is the stage at which you may prefer to do a Southern blot or dot blot analysis of virus DNA (see section 8.2.3) and a protein gel (see section

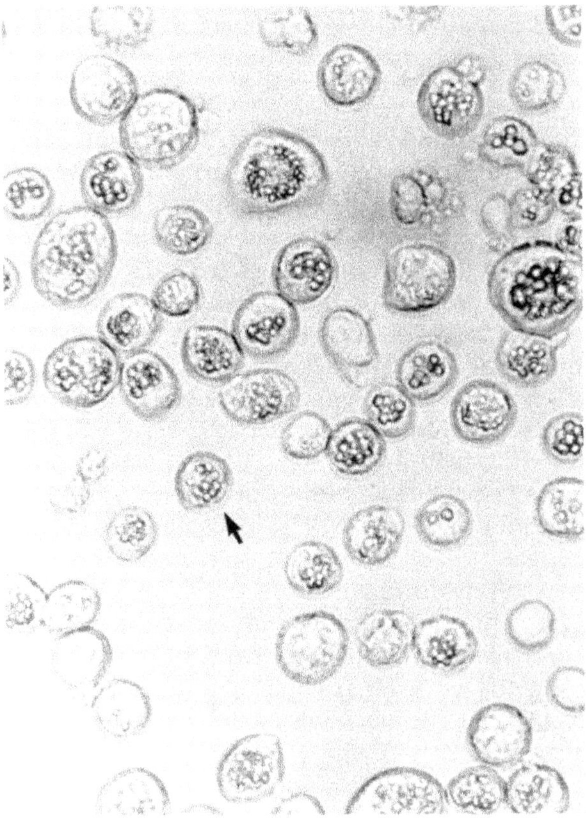

Figure 6.2 Light-microscopy photograph of a plaque showing AcMNPV-infected Sf cells, at 4 days p.i. A typical infected cell showing the presence of polyhedra is indicated by an arrow. We thank Steve Howard for supplying this figure.

8.3.2) to assess incorporation and expression of the foreign gene. However, do not use too much of your precious virus stock on experimental work, prepare intermediate and working stocks for this purpose. If you have not titrated the seed stock, use 100–200 μl to infect 35 mm dishes of cells for the DNA and protein analyses, as described in Chapter 8.

6.5.2 PREPARATION OF AN INTERMEDIATE STOCK OF VIRUS

This section assumes that you are starting with a seed stock of virus, prepared after plaque-purification (section 6.5.1), but applies equally to the amplification of any small stock of virus that you may have been sent as a gift or have purchased. If you prefer, this step may be omitted and you can use your seed stock of virus directly to prepare a large working stock (section 6.5.3).

Method:

1. Seed a 75 cm^2 flask with 5×10^6 Sf cells and infect with seed stock of virus at a low moi (less than 1 pfu/cell if you have titrated the seed stock or use 0.25–0.5 ml if you have not). Remove the medium from the cells and add the required volume of virus inoculum; if the volume is less than 0.5 ml, add medium to increase the volume to this amount. Ensure the inoculum is spread out to cover the complete cell monolayer. Gently rock the inoculum over the cells every 15–20 mins for 1 h.
2. After 1 h at ambient temperature, remove the inoculum and add 10 ml fresh medium. Incubate at 28°C for 4–6 days until the cells are well infected.
   * It is important that a low moi is used at all stages of virus amplification; otherwise it is possible that deletion mutants may occur.
3. Harvest the medium and store at 4°C (or harvest medium and cells for sonication prior to storage; see note in section 6.5.1). This becomes the intermediate stock and can be used for limited experimental work, for example, to test new recombinant viruses for expression. The inoculum should be titrated before use so that you may accurately determine the moi for experimental work (see 6.3.1/2). However, as 10 ml inoculum may not last very long, it is best to amplify a large working stock (50–500 ml) of inoculum as soon as possible.
   * It is worthwhile freezing small amounts of intermediate stock at −70°C, store the remainder at 4°C.

6.5.3 PREPARATION OF A HIGH-TITRE WORKING STOCK OF VIRUS

This section applies to any type of virus (wild-type AcMNPV or recombinant). In the case of new recombinant viruses, amplify only those that you are certain are expressing the gene of interest (use intermediate or seed stocks to ascertain this). If possible use suspension cultures of cells to prepare high-titre working virus stocks (see Chapter 5).

Method:

1. Prepare a small suspension culture of cells at 1×10^5 cells/ml in growth medium (50–500 ml, depending on how many experiments you anticipate carrying out).

* Although suspension cultures will give the highest titre virus stocks, a stock can also be prepared by infecting 5 to 10 × 75 cm^2 flasks, as described above (section 6.5.2).

2. When the suspension culture reaches 5 × 10^5 cells/ml (2–3 days; the cells should not be allowed to overgrow), infect the cells with 0.1–0.2 pfu/cell by simply adding the required volume of intermediate or seed virus stock to the suspension culture. Continue stirring the cells at 28°C for 4–6 days.
3. When the cells are well infected (4–6 days), harvest the medium (the cells can be allowed to settle out from suspension cultures by standing overnight at 4°C), divide into convenient sized aliquots (10 ml) and store at 4°C (keeps at least 1 year). Some aliquots may also be stored at −70°C.
4. Titrate the working inoculum by plaque-assay (see section 6.3). If prepared using suspension cultures the titre of the inoculum should be in the region of 1–2 × 10^8 pfu/ml. If prepared with monolayer cultures it will be lower, about 5 × 10^7 pfu/ml.
   * This step is important if experiments are to be reproducible and if cell infections are to be complete and synchronous.
5. When the working inoculum runs low return to the intermediate or seed stock to prepare a fresh stock. Do not use the remaining working inoculum. If using a batch of virus inocula after a long time period (several weeks/months) it may be best to re-titrate, as the titre will eventually start to drop.

6.6 LARGE-SCALE PRODUCTION OF VIRUS FOR THE PURIFICATION OF VIRUS PARTICLES

This protocol should be followed for the amplification of sufficient quantities of virus (wild-type or recombinant) to allow purification of cell-released virus particles (ECV). Such purification is a pre-requisite to the isolation of virus DNA to use in co-transfections, for example.

Required:

1 l sterile suspension culture flask (see Chapter 5)
400–500 ml TC100 growth medium or alternative (see Chapter 5)
Intermediate or working inoculum of virus to be purified (of known titre)
TE buffer (10 mM Tris-HCl, pH 7.8, 1 mM EDTA)
Sterile, ice-cold 10% (w/v) sucrose in TE buffer
Sterile, ice-cold 50% (w/v) sucrose in TE buffer
Ultracentrifuge tubes for swing-out rotors (14 ml size) and fixed angle-rotor (minimum 50 ml size).

Sterile centrifuge tubes for low-speed centrifugation (2000 r.p.m.)

Method:

1. Seed a suspension culture flask at 1–2 × 10^5 cells/ml in 400–500 ml TC100 growth medium. Grow the cells until the cell density reaches 5 × 10^5 cells/ml.
   * Use cells already adapted to growth in suspension culture (see Chapter 5).
2. Inoculate the cells with the required virus (use Intermediate or working inoculum) at an moi of 0.1–0.2 pfu/cell. Leave the cells stirring for 4–6 days, until all the cells are well infected (inspect a sample under the microscope).
   * Simply calculate the required volume of inoculum required and add to the cells. For example, to infect 500 ml of cells at 5 × 10^5/ml (total = 2.5 × 10^8 cells) with a moi = 0.1, you will need to add 0.25 ml of virus at 1 × 10^8 pfu/ml (total pfu added = 2.5 × 10^7).
3. Working in a safety cabinet preferably, transfer the cells and medium to sterile centrifuge tubes. Remove the cells by centrifugation at 2000–3000 r.p.m. for 10 min at 4°C.
4. Transfer the supernatant to ultracentrifuge tubes and pellet the virus particles at 24 000 r.p.m. (Beckman Ti70 rotor) for 1 h at 4°C. Carefully decant the supernatant into disinfectant. Drain the tubes well and soak the virus pellets in a small volume (total of about 2 ml) TE buffer. Leave the pellets to soak overnight at 4°C. The latter step aids virus resuspension.
5. Resuspend the virus particles carefully by pipetting up and down with a Pasteur pipette. Try not to create excessive froth and aerosols.
6. Layer the virus suspension onto a 10–50%, one-step sucrose gradient in ultracentrifuge tubes suitable for a swing-out rotor.
   * Prepare the gradients by layering 5–7 ml (depending on tube size) 10% (w/v) sucrose in TE buffer (sterile) over a similar volume of 50% (w/v) sucrose. Use the gradients immediately.
7. Band the virus at the interface between the two sucrose solutions by centrifugation at 24 000 r.p.m. (Beckman SW41 rotor) for 1 h in a swing-out rotor at 4°C.
8. The purified virus will be seen as a whitish band at the interface between the two gradients, as illustrated in Figure 6.3. Harvest using a pasteur. Dilute the virus severalfold with TE buffer and concentrate by centrifugation, as in point 4 above. This step removes the sucrose.
9. Soak the virus pellets overnight (2–3 ml TE buffer), resuspend as described above (point 5) and store at 4°C. Do not freeze purified virus.

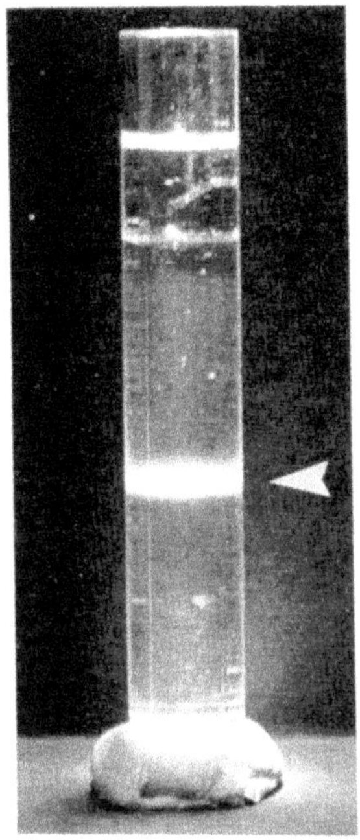

Figure 6.3 An example of a 10–50%, one step sucrose gradient used in the purification of virus particles. The virus particles are banded at the junction between the two sucrose solutions (arrowed).

6.7 PURIFICATION OF INFECTIOUS VIRUS DNA

This method is used to prepare very pure, infectious virus DNA for co-transfections. The starting material is purified virus, either harvested from Sf cells infected *in vitro* (see section 6.6 above) or released from polyhedra, purified from infected larvae (see Chapter 9). The DNA can, of course, also be used for restriction enzyme digests and gel analysis. There is, however, a much quicker method for preparing and analysing virus DNA from infected cells. This protocol is described in Chapter 8.

Required:

400 μl purified virus, prepared as in section 6.6 (or in Chapter 9)
20% (w/v) sarkosyl (Sigma) in 10 mM Tris-HCl, pH 7.8, 10 mM EDTA (sterile)
10 mg/ml ethidium bromide (take appropriate precautions)
50% (w/w) caesium chloride in TE buffer (note w/w and not w/v; i.e. take 50 g caesium chloride and add 50 g (50 ml) TE buffer)
Water bath at 60°C
Ultracentrifuge swing-out rotor and tubes (14 ml size)

Liquid paraffin
Butanol
Ice-cold sterile TE buffer and materials for dialysis (sterile)

Method:
* Remember to use sterile solutions and use aseptic technique throughout, as the DNA you prepare may be used to transfect cells in culture.

1. Take 400 μl purified virus and add 100 μl 20% sarkosyl solution. Incubate at 60°C for 30 min to lyse virions.
2. Immediately transfer the lysed virus sample onto a 5 ml cushion of caesium chloride (50% w/w in TE buffer) containing 12.5 μl ethidium bromide (10 mg/ml). Use a tube designed for swing-out rotors capable of withstanding 35 000 r.p.m. (Beckman SW41 rotor). Top-up the tube with liquid paraffin. Prepare a balance tube or purify a second batch of DNA.
3. Spin at 35000 rpm (Beckman SW41 rotor) for 18 h minimum at 20°C. (Do not spin at 4°C as there is a possibility the caesium chloride may precipitate.)
4. After centrifugation, the virus DNA should be visible as two orange bands without the need to use UV light, as illustrated in Figure 6.4 (try using UV light if you cannot see the bands by eye). The two bands correspond to supercoiled DNA (lower band) and open circular DNA (upper band); harvest both as they are equally infectious. Two methods can be used to harvest the DNA:
   (a) Remove the liquid paraffin and all the caesium chloride solution down as far as the upper band of virus DNA. Change pipettes, and harvest the two virus DNA bands into a sterile bijou. Discard the tube for autoclaving.
   (b) Place the tube in a clamp and retort stand. Bore a fine hole through a rubber bung (to fit the tube that you are using) and place a 21 gauge needle through it; attach a piece of tubing to the luer-lock end of the needle. Place the bung in the top of the tube and clamp the tubing tightly. Using a red hot needle bore a hole in the base of the tube. With a waste container in position to catch the solution as it drips through, slowly release the clamp on the tubing. Collect the DNA fractions in a sterile bijou.
5. Extract the ethidium bromide by adding an equal volume of butanol and shaking gently. Allow phases to separate on the bench (takes 1–2 min) and discard the upper phase containing butanol/ethidium bromide. Dispose of appropriately. Repeat the butanol extraction two or three times until the lower aqueous phase, containing the DNA, is colourless.
6. Place the DNA solution into prepared dialysis tubing and dialyse

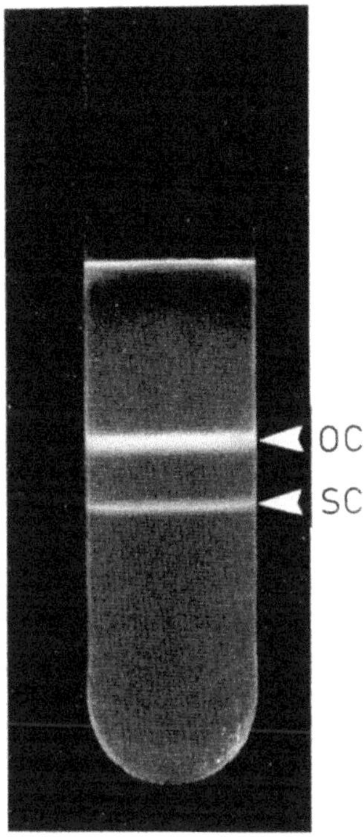

Figure 6.4 An example of a caesium chloride gradient used to purify virus DNA, illustrating the two forms of the virus DNA that should be harvested; super-coiled (SC-DNA) and open-circular (OC-DNA). We thank Martin Ayres for supplying this figure.

against TE buffer (2–3 changes × 500 ml), in a sterile beaker with magnetic stirring bar, at 4°C overnight.

* To prepare dialysis tubing: autoclave or boil narrow bore dialysis tubing and clips in TE buffer for 10 min. Rinse with fresh, sterile TE buffer. Clip one end of the tubing and add the DNA solution. Without trapping air, clip the other end of the tubing. Try to use dialysis clips rather than tie knots in the tubing. Clips are available from most general laboratory suppliers.

7. Following dialysis, transfer the DNA into a bijou and store at 4°C (never store frozen). The DNA appears to remain infectious for up to about 1 year.
8. The concentration of the virus DNA can be determined in a UV spectrophotometer. Take 0.025–0.05 ml of the DNA solution and dilute to 0.5 ml with TE buffer (use a micro-quartz cuvette). Use

TE buffer to zero the spectrophotometer and take an OD reading of the DNA solution at 260 nm. The concentration of DNA (in μg/ml) is 50 × dilution factor × OD_{260} reading.

* As a rule we never ethanol-precipitate the DNA to concentrate it, although we have heard of others who claim to do this without destroying infectivity. The problem lies in resuspending the DNA without it shearing and losing infectivity.

* If you wish to analyse the DNA on a gel, it is best to digest 100–200 ng DNA with the appropriate restriction endonuclease (a good choice is *Hind*III) in a suitable volume of buffer and then concentrate by ethanol precipitation prior to gel-loading, if necessary.

6.8 TITRATION OF VIRUS BY $TCID_{50}$ (END POINT DILUTION)

This is a less satisfactory method, compared to the plaque-assay, of determining the virus titre. The infectious units are expressed as $TCID_{50}$ units/ml and can be converted to an approximate titre in pfu/ml by a simple calculation. The protocol is given here for completeness, but should only really be used for viruses which do not plaque well.

Required:

Sf cells at 2×10^5/ml
96-well cell-culture (microtitre) dishes
TC100 growth medium
Virus to be titred
Discard facilities
Sandwich box and an incubator at 28°C

Method:

1. Make serial log (1 in 10) dilutions of the virus to be titred in TC100 growth medium. You will require 100 μl of each dilution (10 μl/well; 10 wells per dilution). Normally the full range of virus dilutions is plated, from 10^{-1} to 10^{-8}.
2. Take 100 μl diluted virus and add 900 μl Sf cell suspension (2×10^5 cells/ml). Prepare diluted cells in fresh TC100 growth medium. Repeat for each dilution. Mix cell and virus suspension thoroughly, but gently.
3. Add 100 μl cell-virus suspension (10^{-1} dilution) to the first row of wells in a 96-well microtitre plate. The last two wells in the row may have control non-infected cells added. Repeat with the 10^{-2} dilution in the next row and so on, until all the samples have been plated.

4. Incubate the plates at 28°C for 4–6 days, in a box lined with moist tissue paper to prevent drying out. Monitor daily for signs of infection in the lower dilutions.
5. If the virus to titrate is occlusion-positive, leave the cells 4–5 days before scoring: wells are scored for the presence or absence of polyhedra. For occlusion-negative recombinant viruses, leave the cells 5–6 days before scoring because it is more difficult to determine virus-infected cells.
   * If titrating *lac*Z-positive viruses, add 20 μl 2% (w/v) X-gal in DMF to the virus-cell suspension before dispensing into the microtitre dish wells; infected cells will then give rise to blue cells and supernatants.
   * It is important that each well is examined for the presence of virus. Any well that has any infected cells must be scored as positive.
6. Total the number of positive and negative wells for each virus dilution. Calculate the virus titre, as outlined below, using the method of Reed and Muench, as described by Summers and Smith (1987). The $TCID_{50}$ or end point $dilution_{50}$ is determined as the dilution which would give rise to 50% positive wells and 50% negative wells, as extrapolated from the data actually obtained. For the test to be accurate the dilutions used should range from those that give 100% positive and negative wells.
   * An example of how to work out $TCID_{50}$, following the method described by Summers and Smith (1987), is now given:

Raw data (example):

Dilution:	10^{-5}	10^{-6}	10^{-7}	10^{-8}
Positive wells /10:	10	6	1	0
Negative wells /10:	0	4	9	10
Positive total:	17	7	1	0
Negative total:	0	4	13	23
Positive total rate:	17/17	7/11	1/14	0/23
Positive (%):	100	63.6	7.1	0

(a) First calculate the proportionate distance [−PD] (which has a negative value):

−PD= −[(% rate of dilution next above 50%) − 50%] / [(% rate of dilution next above 50%) − (% rate dilution next below 50%)]

In the above example the −PD = −(63.6 −50) / (63.6 −7.1) = −0.24

(b) Calculate log $TCID_{50}$:

log $TCID_{50}$ = log dilution (next to and above 50% positives) + −[PD]

In the example above the log $TCID_{50}$ therefore becomes:

$-6.0 + -0.24 = -6.24$

(c) Convert log $TCID_{50}$ into $TCID_{50}$.

$TCID_{50}$ value $= 10^{\log TCID50}$

For example given $TCID_{50} = 10^{-6.24}$ or $1/(1.74 \times 10^6)$

(d) The titre of the virus in infectious doses per unit of inoculum is obtained from calculating the reciprocal of the $TCID_{50}$ value. The unit dose is normally 1 ml.

In the example given, the titre $= 1.74 \times 10^6$ $TCID_{50}$/0.01 ml (10 μl added per well:

$= 1.74 \times 10^8$ $TCID_{50}$/ml.

(e) To convert $TCID_{50}$ to pfu the following approximate conversion can be used:

$TCID_{50}$/ml × 0.69 pfu/ml

In the example given 1.74×10^8 $TCID_{50}$/ml $\approx 1.2 \times 10^8$ pfu/ml.

Production and selection of recombinant virus 7

7.1 INTRODUCTION

This chapter describes the protocols necessary for the co-transfection of Sf cells with virus DNA and transfer vector, and the subsequent detection, separation and purification of recombinant viruses. Before attempting to make a recombinant virus you should ensure that you: (1) have purified AcMNPV or AcMNPV.*lacZ* DNA, as appropriate (see Chapter 6); (2) have checked that the foreign gene(s) is inserted into the appropriate transfer vector in the correct orientation (see Chapter 4); (3) can carry out a plaque-assay (see Chapter 6).

We will cover two basic techniques that can be used to introduce DNA into insect cells, calcium phosphate co-precipitation (Graham and van der Eb, 1973; Gorman, 1982; Smith and Summers, 1983) and cationic lipid-mediated transfection, or lipofection (Groebe *et al.*, 1990). Electroporation can also be used (Mann and King, 1989), but we will not be mentioning this technique further. Until recently, most workers used the calcium phosphate transfection method, but with the recent introduction of lipofectin (Gibco-BRL), DOTMA (Boehringer Mannheim) and similar reagents from other companies, lipofection is rapidly taking over as the method of choice. It is very easy to carry out, very efficient and gives reproducible results.

One of the main problems that workers new to the baculovirus system face is the detection and separation of recombinant viruses. Selection techniques (discussed in Chapter 2) have involved the tedious and often frustrating process of searching for polyhedrin-negative plaques in a background of polyhedrin-positive plaques. Since the frequency of recombination events are low, this has meant screening up to several hundred plaques to find a few that are recombinant. Another contributory factor has often been the poor quality of the plaque-assays achieved by inexperienced workers. Since these early days, several refinements to the system have been made to make the detection and selection of recombinant viruses easier. These are briefly reviewed below (see Chapter 2 for more information).

(1) *LacZ* selection techniques
By substituting the *lacZ* gene in place of polyhedrin and purifying AcMNPV.*lacZ*, a co-transfection can be carried out with AcMNPV.*lacZ* DNA and the appropriate transfer vector. The mixture of parental (*lacZ*-positive) and recombinant (*lacZ*-negative) viruses can be separated by plaque-assay; staining with X-gal and neutral red; recombinants producing white and parental virus producing blue plaques phenotypes, in a background of pink, non-infected cells.

(2) Using dual-expression vectors to give positive screening techniques
This refinement uses a dual-expression vector (see Chapter 2) in which there is an intact polyhedrin gene, as well as foreign gene, in a co- transfection with AcMNPVpolyhedrin-negative DNA. Parental virus will give rise to polyhedrin-negative plaques and recombinant virus will give rise to polyhedrin-positive plaques.

A similar system in which the polyhedrin gene has been substituted with *lacZ* can also be used, so the parental virus gives rise to white plaques and recombinant virus to blue plaques.

(3) By enriching for recombinant virus prior to plaque-assay
This is achieved by undertaking one or two rounds of limiting dilution infections in microtitre dishes and detecting recombinant virus by dot-blot hybridization, prior to attempting a plaque-assay. This can increase the percentage of recombinants on the first plaque-assay to 30–50%, but it is time consuming (Fung *et al.*, 1988).

(4) By using linearized virus DNA in the co-transfection
Perhaps the most significant advance in both speeding-up and making the selection process easier is the use of linearized virus DNA in the co-transfection with the transfer vector (Kitts *et al.*, 1990). The theory behind this advance has been discussed in Chapter 2. There are two viruses currently available that can be linearized, AcMNPV.SC and AcMNPV.*lacZ*. In the former a unique *Sau*I (*Mst*II or *Bsu*36I) site has been introduced into the polyhedrin gene and in the latter the *lacZ* gene itself fortuitously contains a *Sau*I site. Using linearized DNA from either virus in a co-transfection will give rise to 20–40% recombinants (or more) in the first round of screening by plaque-assay. If you use AcMNPV.SC there will be no visible difference between parental and recombinant plaques (both are polyhedrin-negative), however, since up to four or more plaques out of ten will be recombinant, you simply need to pick

up to ten plaques for analysis by dot-blot hybridization or PCR techniques. If you are using linearized AcMNPV.*lacZ*, the process is even easier since parental virus will give rise to blue plaques and recombinants to white plaques upon staining with X-gal and neutral red. The number of false-recombinants when using the *lacZ*-selection system (presumably due to spontaneous mutations disrupting the *lacZ* coding sequence) can vary from less than 10 to about 30% of putative recombinants, after the first round of plaque-assay. It is therefore important that you take five or six different plaque-picks through to the second and third rounds of plaque-purification and amplification. Alternatively, you could use PCR techniques (see Innis *et al.*, 1990) to detect the presence of authentic recombinants after the first round of plaque-purification, thus saving time and effort in plaque-purifying viruses which are false-recombinants.

We strongly recommend that you generate your recombinant virus using linearized virus DNA, together with the lipofection co-transfection technique. AcMNPV.SC or AcMNPV.*lacZ* can be obtained on request from IVEM, Oxford. In this way you will rapidly produce and purify your recombinant virus and avoid much of the frustration and problems mentioned at the beginning of this chapter. We suggest that you read through the flow-chart in Figure 7.1, which outlines the pathway to producing recombinant virus ready for analysis of expression (see Chapter 8), before attempting your first co-transfection.

7.2 PREPARATION OF LINEAR AcMNPV.*lacZ* (or AcMNPV.SC) DNA

Refer to Chapter 6 for the techniques needed to amplify and purify virus, and subsequently virus DNA, from Sf cells.

Required:

AcMNPV.SC or AcMNPV.*lacZ* DNA of known concentration
10× RE buffer (usually supplied by the manufacturer)
*Sau*I RE (or *Mst*II, *Bsu*36I)
Sterile water
Water baths at 37°C and 60°C

Method:

1. It is convenient to digest 2 μg virus DNA at one time. Set up the digest in the usual way, for example:

AcMNPV.SC or .*lacZ* DNA	2 μg
10× RE buffer	10 μl
*Sau*I RE	10 units
Water	to 100 μl

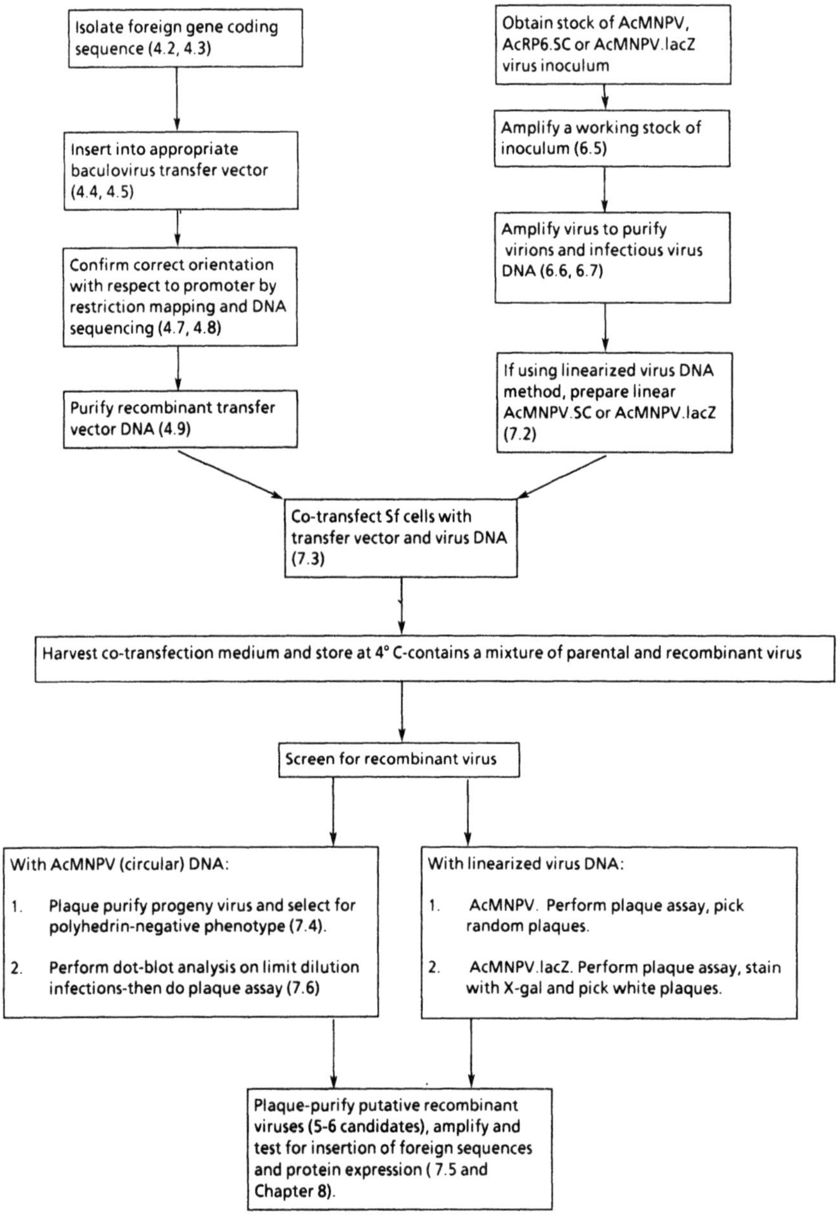

Figure 7.1 A guide to the steps that need to be taken in the production, detection and amplification of a recombinant baculovirus.

Incubate at 37°C for 4–6 h (or can be left overnight).
* Remember to keep the DNA sterile, as you will be using it to transfect cells in culture. Kitts *et al.* (1990) demonstrated that the highest percentage of recombinant virus was obtained when the virus DNA was over-digested with excess RE with a long incubation time.

2. Remove a 100 ng sample for analysis in a 0.6% agarose mini-gel to check that the DNA has been linearized. You will need to run a 100 ng sample of non-digested virus DNA in a parallel track as a control. Store the digested DNA on ice during this procedure. If the DNA has not digested, add more enzyme and continue the incubation for a few more hours. An example of gel showing linearized and circular virus DNA is shown in Figure 7.2.
3. Heat-inactivate the RE by incubating at 60–65°C for 5 min. Spin the sample briefly in a microfuge and store the DNA at 4°C. You require 100–200 ng DNA for each co-transfection using

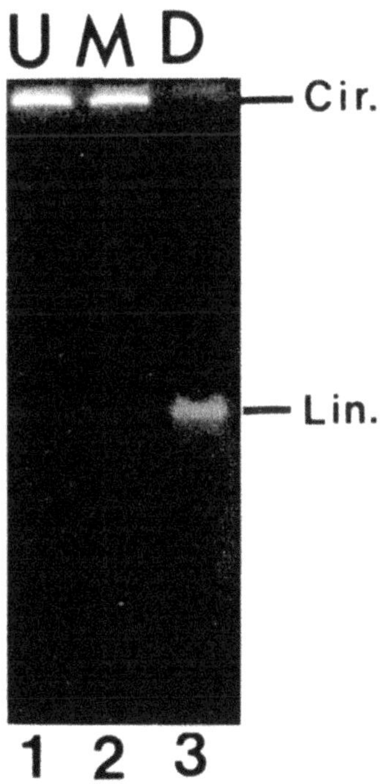

Figure 7.2 This 0.6% agarose mini-gel illustrates the different migration rates of undigested (U, track 1) mock-digested (M, track 2) and linearized (D, track 3) AcMNPV.SC DNA. The DNA was linearized with *Bsu*36I. The circular (Cir.) and linear (Lin.) forms of the virus DNA are indicated. Reproduced from Kitts *et al.* (1990) with permission.

lipofection and 1 μg DNA if using the calcium phosphate method.

7.3 CO-TRANSFECTION OF INSECT CELLS

Methods for lipofection and calcium phosphate transfection of Sf cells are included, although we strongly recommend that you use the former technique in conjunction with linearized virus DNA.

7.3.1 CO-TRANSFECTION USING LIPOFECTION

Required:

35 mm dish of Sf cells seeded with 1×10^6 cells (see Chapter 5)
TC100 medium without serum added (with antibiotics)
TC100/10% FCS growth medium
Transfer vector (see Chapter 4)
Linearized virus DNA (see section 7.2) or intact virus DNA (see section 6.7)
Lipofectin (Gibco-BRL) or DOTMA (Boehringer Mannheim) or similar reagent
Polystyrene (not polypropylene) container
Sandwich box lined with moist tissue paper

Method:

1. For each co-transfection you will require 1 × 35 mm dish seeded with 10^6 Sf cells, prepared as described in Chapter 5. You may prefer to carry out the co-transfection in duplicate.
2. Mix together the virus and transfer vector DNAs in a polystyrene container (a disposable universal bottle is useful, as the small drop of liquid falls into the conical point). Make a note of the combined volume of the DNA solutions.
   * If using linearized DNA, you will need 100–200 ng virus DNA and 500–1000 ng (5× excess by weight) transfer vector.
   * If using intact virus DNA, you will require about 500 ng virus DNA and 2.5 μg transfer vector.
3. Add an equal volume of diluted lipofectin to the DNA solution in the universal bottle. Lipofectin must be diluted before use by adding 2 parts lipofectin to 1 part sterile water.
   * For example, if the combined DNA volumes are 24 μl you will require 24 μl diluted lipofectin (16 μl lipofectin:8 μl sterile water).
4. Mix the diluted lipid-reagent with the DNA solution and leave at room temperature for 15 min, to allow the liposome–DNA complexes to form.

5. Meanwhile, remove the medium from the dish(es) of Sf cells and wash the monolayer twice with TC100 medium without serum. To do this, very gently pipette 1–2 ml medium onto the cells, rock the dish very gently once or twice and discard the medium. Repeat.
6. Place 1 ml TC100 medium without serum into the dish of cells. When ready, add the DNA-lipid mixture and swirl gently to mix in. Incubate the cells in a humidified atmosphere at 28°C for 5 h, or leave overnight.
7. After 5 h (or overnight) add an extra 1 ml TC100/10% FCS to the dish(es) of cells and continue the incubation until 48 h post-transfection.
8. Harvest the 2 ml medium into a sterile container, for example a disposable bijou, and store at 4°C. This medium contains a mixture of parental and recombinant virus.
   * The next step is to separate the two virus types by plaque-assay (section 7.4). If you have not used the linear-DNA method, you may prefer to include a round of amplification and dot-blot hybridization before attempting a plaque-assay (see section 7.6)
   * If you want to check that the co-transfection has worked at this stage and you have used a *lac*Z-positive parental virus, simply add 1–2 ml TC100 growth medium containing 12.5–25 μl (2% w/v) X-gal to the remaining monolayer of transfected cells. Within a few hours at 28°C, the cells and medium should start to turn blue if the co-transfection has been successful. If the parental virus was polyhedrin-positive, you can simply add more TC100 growth medium to the monolayer of cells and inspect them daily for the appearance of polyhedra. If the co-transfection has worked, polyhedra can easily be seen in the cell nuclei 3–4 days post-transfection.

7.3.2 CO-TRANSFECTION BY CALCIUM PHOSPHATE CO-PRECIPITATION

An important technical point to be mentioned with this method is that success very much depends on forming the correct size of precipitate: too small and the cells will not be stimulated to take up the DNA, too large and clumpy and the again the cells will not be able to take up the DNA. The size of the precipitate is dependent on the pH of the medium. An optimal precipitate is formed at a pH of 7.05 (Gorman, 1982); below this pH the precipitate formed may be too fine, or may not form at all, and above this pH the precipitate formed may be too coarse or clumpy.

Required:

35 mm dish seeded with 1 × 10^6 Sf cells
TC100 growth medium with antibiotics
Virus DNA (linearized as in section 7.2 or intact, see section 6.7)
Transfer vector (see Chapter 4)
Sterile water
Filter-sterilized 2 M $CaCl_2$
Filter-sterilized 100 mM glucose
Filter-sterilized 2× HEPES pH 7.05 (40 mM HEPES, pH 7.05, 2 mM Na_2HPO_4, 10 mM KCl, 280 mM NaCl)
Sandwich box lined with moist tissue paper
Whirly-mixer
Deep sterile tube (e.g. disposable universal, centrifuge tube or 5 ml cryogenic vial)

Method:

1. Prepare a 35 mm dish (or two if you prefer to do a duplicate) of Sf cells for each co-transfection, as described in Chapter 5.
2. Set up the co-transfection mix in a deep, sterile tube as follows:

Virus DNA (linear or intact)	1 μg
Transfer vector	5 μg
2× HEPES buffer	500 μl
100 mM glucose	100 μl
Water	to 935 μl

3. Whilst vortexing the mixture, add 65 μl 2 M $CaCl_2$ in a dropwise manner.
4. Leave the mixture at room temperature for 30 min for the precipitate to form.
5. Remove the medium from the dish(es) of cells and replace with the 1 ml of co-transfection mixture. Incubate the cells at room temperature for 30 min.
6. Without removing the co-transfection solution, add 1 ml TC100 growth medium to the dish(es) of cells and incubate in an humidified atmosphere at 28°C for 2–4 h.
   * After this incubation, the precipitate should have fallen onto the cells and be visible, as a fine layer, under the light microscope.
7. After the 2–4 h incubation, discard the medium and replace with 2 ml fresh TC100 growth medium. Continue to incubate the cells at 28°C until 48 h post-transfection.
   * If the precipitate is slightly too dense, it may be worthwhile washing the monolayer once or twice, to remove excess precipitate, before the final incubation

8. Harvest the medium, containing a mixture of parental and recombinant virus types, and store at 4°C.
   * The next step is to separate the two virus types, either by plaque-assay (see section 7.4) or by amplification and dot-blot hybridization (section 7.6) prior to plaque-assay.
   * See note to section 7.3.1, point 8, for details of how to ascertain at this point whether or not the co-transfection has worked. Bear in mind that the cells will be in a relatively poor condition, compared with those treated with lipofectin, and that it will not be so easy to distinguish polyhedra.

7.4 SEPARATION OF PARENTAL AND RECOMBINANT VIRUSES BY PLAQUE-ASSAY

A detailed protocol for performing a plaque-assay has been described in Chapter 6, section 3. Here, we present a shortened protocol, but one that is directed at using a plaque-assay to distinguish recombinant and parental virus types. If your starting material is medium harvested from a co-transfection using linearized virus DNA, you should not have any problems in finding recombinant virus after screening relatively few plaques. If you have used the AcMNPV.SC DNA then you will need to pick about ten plaques 'blind' for analysis by dot-blot hybridization or PCR techniques to confirm the identity of the recombinant plaques. If you have not used linear-DNA in the co-transfection you can still screen for recombinants by direct plaque-assay, you will simply need to screen more plaques. Alternatively, you may prefer to amplify the recombinant first (see section 7.6) and then do a plaque-assay.

Required:

35 mm dishes of Sf cells, prepared for plaque-assay as described in section 6.3.
TC100 growth medium with antibiotics
2% (w/v) low-gelling temperature agarose (sterilized by autocaving, see section 6.3), cooled and mixed with an equal volume of TC100 growth medium (hold at 37°C until ready)
Harvested co-transfection medium (see section 7.3)
0.5% (w/v) neutral red (filter-sterilized)
PBS
2% (w/v) X-gal in DMF (if using *lac*Z selection)
Water bath at 37°C
Sandwich box lined with moist tissue paper

In our experience Sf21 cells give more well-defined plaques than Sf9 cells, thus making the screening process for recombinant viruses easier.

Method:

1. Prepare the required number of 35 mm dishes of Sf cells. This will depend on the co-transfection method used and the number of dilutions to be assayed. The following gives some guidelines.

   Linear DNA method (lipofection/calcium phosphate): three dishes of cells for each dilution to be assayed (assay neat, 10^{-1}, 10^{-2} and 10^{-3} dilutions).

   Non-linear DNA method (lipofection or calcium phosphate): ten dishes of cells for each dilution to be assayed (assay 10^{-1}, 10^{-2}, 10^{-3} and 10^{-4} dilutions)

   Plaque-assay following amplification/dot-blot screening: five dishes each at 10^{-2}, 10^{-3} and 10^{-4} dilutions.
2. Prepare the required dilutions of the saved co-transfection medium in TC100 growth medium. For each dilution you will require 100 μl for each dish of cells to be infected. Use sterile microfuge tubes or bijoux for the dilutions. Unless using immediately, store virus dilutions on ice.
3. Remove the medium from the Sf cells and add 100 μl of the appropriate virus dilution. Leave for 1 h at room temperature.
4. Discard the virus inoculum into disinfectant and cover the cells with 2 ml agarose/TC100 solid overlay (see Chapter 6, section 3 for more details).
5. When the agarose has set, overlay each dish with 1 ml TC100 growth medium (liquid overlay).
6. Place the dishes of cells into a sandwich box and incubate at 28°C for 3–4 days.
7. Stain the dishes of cells to visualize the plaques. The staining method will depend on the selection regime you have chosen. For more detailed information on staining see Chapter 6. As a guide.

   To distinguish polyhedra-negative from polyhedra-positive plaques: stain with neutral red alone. When the plaques have cleared, examine under the ×10 objective of the inverted microscope and ring several plaques that you think are polyhedra-negative. Examine these in more detail using the ×40 objective. Figure 7.3 illustrates the differences between polyhedra-negative and polyhedra-positive plaques.

   To distinguish *lac*Z-negative from *lac*Z-positive plaques and where you want to pick the latter: stain with X-gal only. The *lac*Z-positive plaques will turn blue.

   As above, but where you want to pick the *lac*Z-negative plaques: dual stain with X-gal and neutral red. The *lac*Z-negative plaques will form clear patches in a background of pink-stained, non-infected cells.

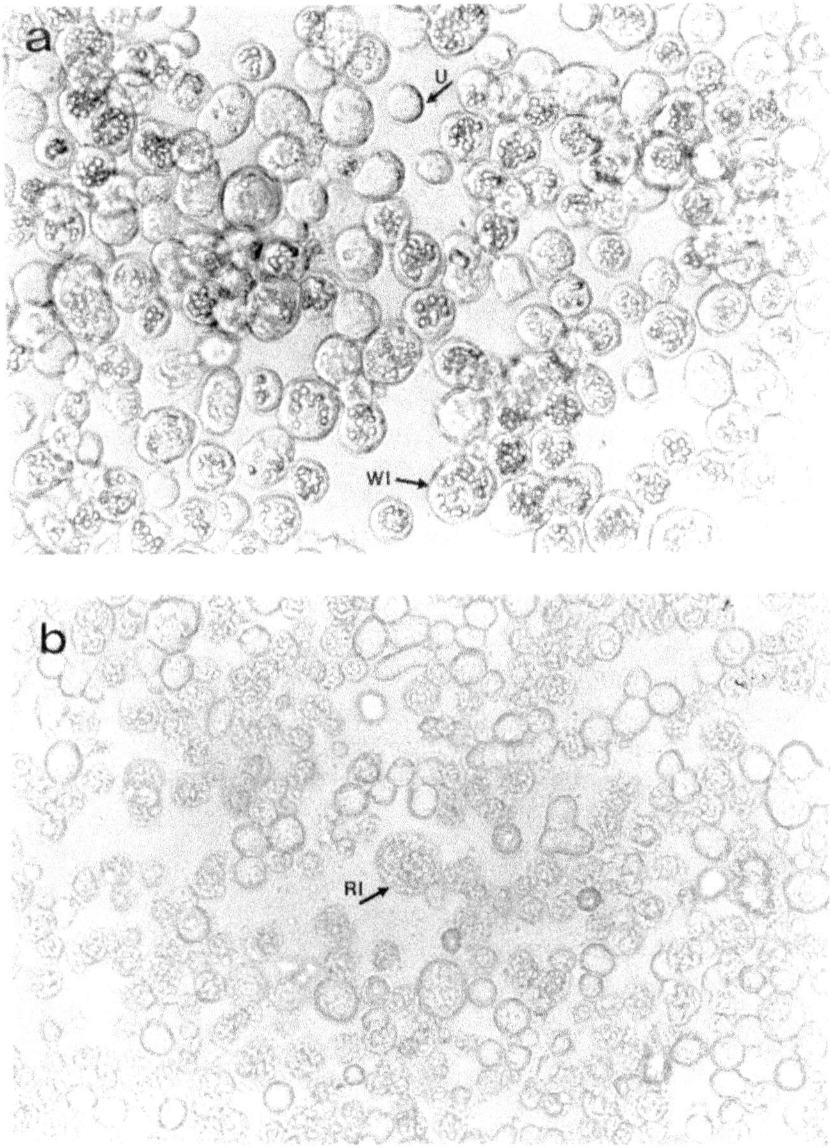

Figure 7.3 The photographs illustrate the differences, as observed using a ×40 objective on an inverted microscope, between polyhedra-positive (panel a) and polyhedra-negative (panel b) plaques. Wild type AcMNPV- (WI), recombinant virus- (RI) and uninfected (U) Sf21 cells are indicated by arrows. We thank Steve Howard for supplying this figure.

8. Having identified a number of putative recombinant plaques, the next step is plaque-purification, followed by the amplification of virus stocks.

7.5 PLAQUE-PURIFICATION AND AMPLIFICATION OF RECOMBINANT VIRUS STOCKS

If facilities for PCR are available, it is possible to use this technique to confirm the identity of a putative recombinant virus after the first plaque-pick. Simply use 1 μl of the TC100 medium containing the plaque-picked virus directly in a PCR reaction, using the appropriate primers for the foreign gene sequences (Malcolm Banks and Louis Obosi, personal communications). In this way you need only further plaque purify those viruses that definitely contain the foreign gene.

7.6 AMPLIFICATION AND DETECTION OF RECOMBINANT VIRUSES BY LIMITING DILUTION AND DOT-BLOT HYBRIDIZATION

If you are not using the linearized-DNA method of co-transfection and are finding it difficult to detect recombinant virus plaques, then you may prefer to amplify the recombinant virus by limited-dilution infections of Sf cells in microtitre dishes. This is followed by screening for recombinants by dot-blot hybridization or by using PCR techniques.

Required:

Sf cells at 1.5×10^5 cells/ml in fresh TC100 growth medium (see Chapter 5 for more details)
96-well tissue culture (microtitre) dishes
Harvested co-transfection medium for analysis
TC100 growth medium
Sandwich box lined with moist tissue paper
Multi-channel pipettor (optional)
0.2 M NaOH
20× SSC (see Chapter 8)
2 M NaCl
Hybridization membrane (e.g. Hybond from Amersham International)
3 MM filter paper
Dot-blotting apparatus
Materials for making a radiolabelled probe (see Sambrook *et al.*, 1989)
Materials for hybridization (see Chapter 8)
Oven at 80°C

Method:

1. Make serial log (1 in 10) dilutions of the harvested co-transfection medium from 10^{-1} to 10^{-4}. You will require 1 ml of each dilution. Make dilutions using TC100 growth medium.
2. Place 100 μl Sf cells (1.5×10^5 cells/ml) into the wells of a 96-well dish. You will require approximately 10 ml of cells.
3. Using 10 wells in each row, add 50 μl of the 10^{-1} dilution to rows 1 and 2. Add 50 μl of the 10^{-2} dilution to rows 3 and 4, and so on.
4. The unused wells at the ends of each row can be used as controls, e.g. for non-infected and wild-type virus infected cells.
5. Place the dish in a moist sandwich box and incubate at 28°C for 4–5 days, or until the cells appear well infected with virus.
6. Transfer the culture medium from each well into a second microtitre dish (the use of a multi-channel pipettor is recommended), seal the dish with parafilm and store at 4°C.
7. Lyse the cells remaining in the original dish by adding 200 μl 0.2 M NaOH. Scrape the cells into the solution and leave for 15 min at room temperature (use tips attached to a multi-channel pipettor as scrapers).
8. Set up the dot-blot filtration apparatus according to the manufacturers' instructions. It is important that the sheets of 3 MM paper and hybridization membrane used are soaked in 20× SSC buffer before placing in the apparatus. Filter 400 μl 2 M NaCl through each well, followed by the 200 μl of cell lysate.

   * It is worthwhile using one or more of the 'spare' wells to include a positive control for the hybridization probe. For example, add 50 ng of the transfer vector DNA that contains your foreign gene of interest. This DNA must be denatured prior to adding to the dot blot (treat with 0.2 M NaOH or boil for 2 min).
9. Bake the filter at 80°C for 2 h and then hybridize with a radiolabelled probe (specific for the foreign gene), as described in Chapter 8 or according to your own protocol. Wash the filter sequentially, at room temperature, with 2× SSC, 0.1% SDS (15 min), 0.5× SSC, 0.1% SDS (15 min), 0.1× SSC, 0.1% SDS (15 min) and at 50°C with 0.1× SSC, 0.1% SDS (2 × 15 min). Wrap the membrane in saran-wrap or cling-film and expose to X-ray film to obtain an autoradiograph. Often the signal from the lysed cell samples are weak and you may need to expose the X-ray film for several days. Figure 7.4 illustrates a typical dot-blot screening result.

   * Think about how you are going to orientate the autoradiograph with the stored microtitre dish i.e. mark the filter with radioactive ink.

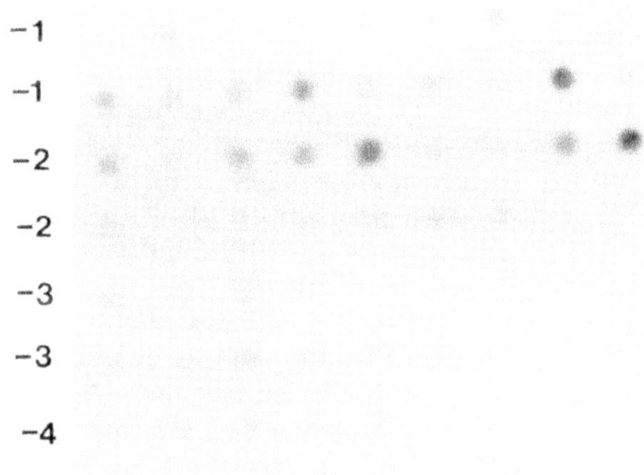

Figure 7.4 A typical autoradiograph obtained when screening for recombinant virus after limited dilution infections in microtitre dishes. The log dilutions of the co-transfection medium used to infect the cells in the microtitre-dish are shown on the left-hand side. We thank Allan Atkinson for supplying this figure.

10. Orientate the autoradiograph with the stored microtitre dish and identify the wells that contain recombinant virus. Any positive wells from the 10^{-3} and 10^{-4} dilution rows will contain a higher percentage of recombinant virus than the positive wells from the other rows.
11. The next step is to set up a plaque-assay, using the culture medium in the positive wells of the stored dish as the source of inoculum Take 50 μl of the required medium and add to 450 μl TC100 growth medium to prepare the 10^{-1} dilution. Continue making 10^{-2}, 10^{-3} and 10^{-4} dilutions. Carry out a plaque-assay as described in section 7.4 and in Chapter 6. Having identified recombinant plaques, you can then proceed to plaque-purify and amplify stocks of recombinant virus.

Characterization of recombinant viruses 8

8.1 INTRODUCTION

The most common practice for workers who have just isolated a putative recombinant baculovirus is immediately to test for the expression of the foreign gene using an appropriate assay for the protein. This is inevitable, but not the correct procedure to adopt. Instead, one should analyse the genome of the recombinant baculovirus to ensure that the foreign coding sequences have been inserted correctly in the desired position before monitoring protein production. While it is relatively uncommon for there to be a problem in the recombination process, we have noted occasions when viruses were polyhedrin- or *lac*Z-negative but lacked the foreign coding sequences. If this is determined before extensive tests are performed for protein production, then considerable disappointment can be avoided. Understandably, very few of us have the patience to do things in the correct order! For a compromise it is best to investigate protein synthesis at the same time as analysing the recombinant virus genome.

Another important point to consider is whether the recombinant virus stock has been accurately titred. If the infectivity of the stock(s) is too low then the insect cells will be inadequately infected and virus gene expression will be poor. In some cases we have encountered, workers have used virus stocks prior to infectivity titrations and have been disappointed when little protein production has been observed. Subsequent plaque-assays have allowed the titre of the virus stocks to be determined, thus permitting an accurate moi to be used in tests for gene expression. This has usually produced results which have been received with considerably more enthusiasm!

The protocols described below outline the procedures for critically analysing the genome of a recombinant virus (Southern hybridization); testing the level of protein synthesis in infected cells

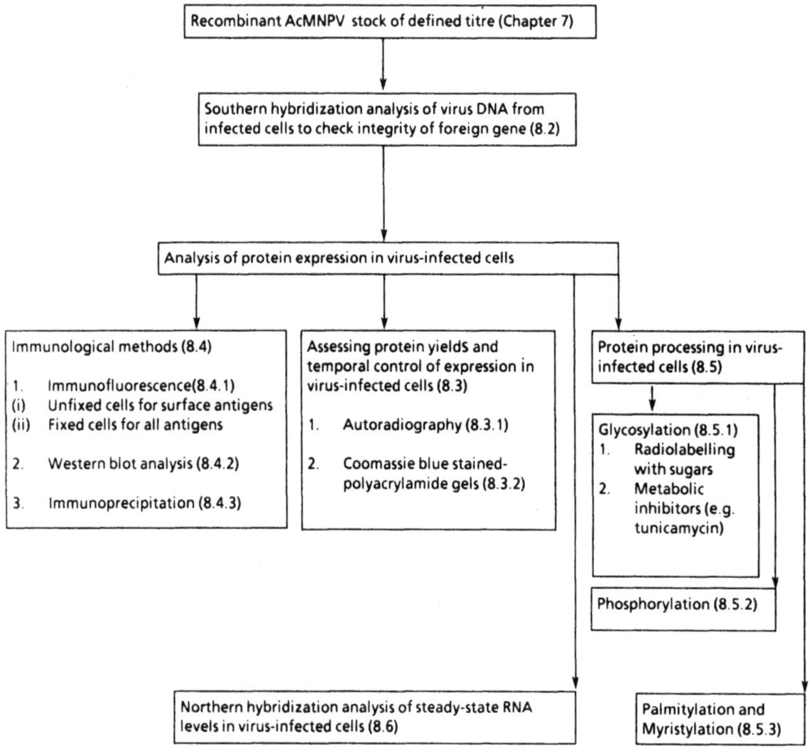

Figure 8.1 Flow-chart illustrating the techniques that can be used to analyse recombinant baculoviruses.

(radiolabelling of cells; PAGE; Western blots; immunofluorescence); monitoring the steady state levels of mRNA (Northern hybridization). Figure 8.1 shows a flow-diagram outlining the various methods that can be used to analyse a recombinant virus.

8.2 ANALYSIS OF RECOMBINANT VIRUS GENOMES

In Chapter 7, a preliminary test for determining whether the putative recombinant virus had the foreign gene inserted was described (dot-blot hybridization). This is done at a very early stage in the isolation procedure. In this section, we describe the procedure for analysing the virus genome once the final stock of recombinant virus is available.

8.2.1 EXTRACTION OF DNA FROM VIRUS-INFECTED CELLS

Required:

1 × 35 mm dish of Sf cells for each virus to be analysed
2 × 35 mm dishes of Sf cells for controls (wild type AcMNPV and mock-infected cells)
TC100 growth medium
Virus stocks for inoculation (preferably of known titre)
Phosphate buffered saline (PBS)
TE buffer (10 mM Tris-HCl, pH 8.0, 1mM EDTA)
Cell lysis buffer (50 mM Tris-HCl, pH 8.0, 5% 2-mercaptoethanol, 0.4 % w/v SDS, 10 mM EDTA)
Proteinase K (10 mg/ml in TE; make up fresh each time and preincubate at 37°C for 30 min prior to use)
Ribonuclease A (10 mg/ml; heated to 100°C for 10 min prior to use)
Phenol/chloroform (50:50, equilibrated in 50 mM Tris-HCl, pH 8.0)
3 M sodium acetate
75% and 100% ethanol

Method:

1. Check that the cells have formed a sub-confluent monolayer. Remove the medium and inoculate cells with 10 pfu virus/cell (recombinants and a wild-type AcMNPV control). Mock-infect one dish with TC100 medium (100 μl). Leave the cells at room temperature for 1 h.
   * See Chapters 5 and 6 for more details on setting up cells and infecting them with virus.
2. Remove the inoculum and discard into disinfectant. Overlay the cells carefully with 1.5 ml TC100/FCS growth medium. Incubate the plates in a moist sandwich box for 18 h at 28°C.
3. Harvest the cells with a Pasteur pipette and place in a 1.5 ml microfuge tube.
   * When the cells have been infected with virus for 18 h or longer they will readily detach from the plastic surface when medium is aspirated over them using the Pasteur pipette. Uninfected cells usually require scraping off with a rubber policeman.
4. Pellet the cells using low-speed centrifugation and discard the medium disinfectant. Wash the pellet once with 0.5 ml PBS.
5. Resuspend the infected cells in 250 μl TE. Add 250 μl cell lysis buffer and gently mix the contents of the tube. The lysate should become sticky and viscous as the cells release DNA.
6. Add 12.5 μl proteinase K and 2.5 μl ribonuclease A and incubate at 37°C for 30 min.

* The protease might be expected to inactivate the ribonuclease; in practice the concentration of the latter enzyme is so high that RNA is satisfactorily degraded.

7. Extract the lysate with an equal volume of phenol/chloroform for 5 min. Ensure that the organic and aqueous phases are thoroughly mixed, but avoid violent shaking as this shears the high molecular-weight virus DNA. Separate the phases by spinning for 2 min in a microfuge.
8. Remove the aqueous phase and place in fresh tube. It is sometimes difficult to avoid removing some of the interface at this point. This may be largely avoided if a wide bore Pasteur pipette is used or a 1 ml Gilson tip with the narrow bore removed.
9. Repeat steps 7 and 8.
10. Add 50 μl 3 M sodium acetate and two volumes of ethanol to the aqueous phase to precipitate the DNA.
    * A precipitate is usually observed immediately; there is no need to chill the tubes.
11. Pellet the DNA for 5 min in a microfuge and wash the pellet twice with 75% ethanol. Remove the last traces of ethanol with a drawn-out Pasteur pipette.
12. Air dry the DNA and add 100 μl TE. Do not attempt to resuspend the DNA at this point; soak it overnight at 4°C.
13. Incubate the DNA at 37°C for 10 min and then gently resuspend the DNA with a Gilson tip.
14. Store the DNA at 4°C.

8.2.2 ANALYSIS OF VIRUS DNA WITH RESTRICTION ENDONUCLEASES

The virus DNA prepared using the method described above (section 8.2.1) is suitable for dot blot and Southern hybridization analysis. The yield of DNA from 10^6 Sf cells is quite high (>5 μg) and there should be sufficient material for several experiments. When the DNA is digested with restriction endonucleases (RE) and analysed with an ethidium bromide-stained agarose gel, the clarity of the profile is often surprisingly good. Very little chromosomal material obscures the virus DNA and, if the digest is appropriate, it is possible to identify the inserted DNA sequences when compared with the wild-type control. An example of such an analysis is shown in Figure 8.2.

The choice of RE to use will depend on the transfer vector and inserted gene sequences, ideally you would choose an enzyme that excises the foreign gene insert so that an extra band appears in the RE profile of the recombinant virus compared with wild-type AcMNPV

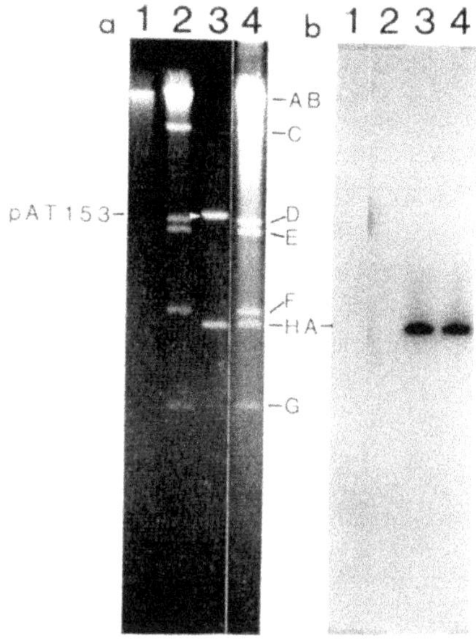

Figure 8.2 Southern hybridization of DNA from virus-infected cells. Panel a: ethidium bromide-stained 0.6% agarose gel of *Bam*HI-digested DNA. Lane 1, uninfected Sf cells; lane 2, AcMNPV DNA from purified virions; lane 3, pAT153 containing the influenza virus HA gene; recombinant virus, AcRP5.HA-infected Sf cells. Panel b: hybridization analysis of the agarose gel shown in panel a using nick-translated HA gene fragment as a probe. Reproduced from Possee (1986) with permission.

(for an example, see Figure 8.2). If this is not possible, then a *Bam*HI or *Hin*dIII digest of the virus DNAs will give suitable profiles for analysis by Southern blotting.

Required:

10× RE buffer, as required
RE as required
5× DNA sample mix (5× TBE (see section 4.2.2), 25% sucrose or glycerol, 2.5% SDS, trace of bromophenol blue)
Materials for agarose mini-gel electrophoresis
DNA size markers (e.g. λ DNA digested with *Eco*RI and *Hin*dIII).
Virus DNA (prepared as in section 8.2.1)

Method:

1. Set up the following RE digests for each DNA sample:

DNA	5–10 μl
10× RE buffer	2 μl
RE (5 units)	0.5 μl
Water to a final volume of 20 μl	

* The volume of the reaction mix may be varied according to the type of agarose gel to be used.

* Controls should include non-infected Sf cell DNA, wild-type AcMNPV-infected Sf cell DNA and appropriate DNA size markers.

2. Incubate for 2 h at 37°C then add 5 μl of 5× DNA sample mix.
3. Analyse by electrophoresis in a 0.6–0.8% agarose gel and stain with ethidium bromide.

   * The concentration of agarose in the gel is determined by the size of the DNA fragments to be resolved.
4. Photograph the gel and proceed to section 8.2.3 for a description of the Southern hybridization method, if necessary.

8.2.3 SOUTHERN HYBRIDIZATION ANALYSIS OF VIRUS DNA

Required:

Denaturing solution (1.5 M NaCl, 0.5 M NaOH)
Neutralizing solution (3.0 M NaCl, 0.5 M Tris-HCl, pH 7.0)
20× SSC (3.0 M NaCl, 0.3 M tri-sodium citrate)
2× SSC (diluted from 20× SSC stock)
Hybridization solution (50% deionized formamide [stir with Amberlite MB-3 (Sigma) to pH 7.0, then filter], 3× SSC, 50 mM HEPES, pH 7.0, 0.1% polyvinylpyrrolidone, 0.1% ficoll, 0.1% bovine serum albumin)
Hybridization probe: [^{32}P]-DNA (to give not greater than 5 ng/ml in the hybridization solution)
Sonicated salmon testes DNA (sufficient to give a final concentration of 10 μg/ml in the hybridization buffer)
Hybridization membrane (Nitrocellulose or equivalent, e.g. Hybond N from Amersham).
Whatman 3MM paper
DNA transfer apparatus (see Figure 8.3)
Agarose gel with virus DNA (from section 8.2.2)

Method:

1. Soak the agarose gel (see section 8.2.2) in denaturing solution for 1 h (20 × 20 cm gel) or 15 min (mini-gel).
2. Remove the denaturing solution and rinse gel with a small volume of neutralizing solution. Soak the gel in neutralizing solution for 1 h (20 × 20 cm gel) or 15 min (mini-gel).
3. Place gel on transfer apparatus and assemble as depicted in Figure 8.3. Transfer DNA to membrane, overnight for a 20 × 20 cm gel or 2–3 h for a mini-gel.
4. Dismantle the transfer apparatus. Mark positions of sample wells

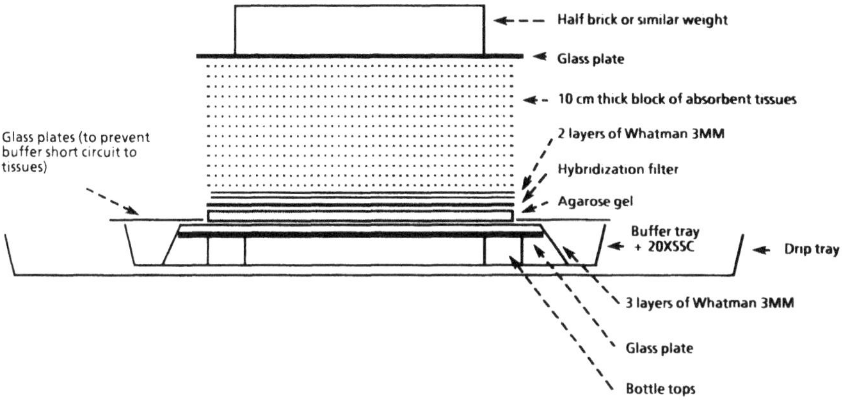

Figure 8.3 Apparatus used for the transfer of DNA (or RNA) from an agarose gel to a hybridization membrane.

on hybridization membrane. Do not wash the membrane. Discard the gel and bake the membrane for 2 h at 80°C.

5. Soak the membrane in sterile water for 15 min to soften salt deposits (optional).
6. Boil the membrane in 1 mM EDTA for 5 min; use a microwave oven or pyrex/enamel dish on a hot-plate.
   * This procedure replaces the conventional pre-hybridization step and is very effective at lowering non-specific binding of radioactive probe to the membrane, and saving time.
7. Place filter in bag and seal on three sides, or place in a commercial hybridization-bottle. Boil the radioactive probe and salmon testes DNA for 3 min to denature the DNA and then snap chill on dry-ice/ethanol (if available). Add the denatured DNA to the hybridization solution and place in the bag or bottle.
8. Incubate at 42°C for 16 h with constant agitation.
9. Remove membrane from the bag (or bottle) and wash four times with 2× SSC/0.1% SDS at room temperature (briefly) and twice with 2× SSC/0.1% SDS at 65°C (30 min each wash).
10. Seal the membrane in a plastic bag immediately to avoid drying.
    * If the membrane is exposed to X-ray film while it is still wet then the probe may be readily removed (5 min boiling in 1 mM EDTA) prior to rehybridization.
11. Expose to X-ray film. A 6–12 h exposure is usually sufficient to detect the desired DNA bands.

8.3 ANALYSIS OF FOREIGN GENE EXPRESSION BY POLYACRYLAMIDE GEL ELECTROPHORESIS, USING UNLABELLED OR RADIOLABELLED CELL EXTRACTS

These are the most commonly used and convenient procedures for analysing the level of synthesis of the foreign protein in recombinant virus-infected insect cells. More sensitive methods, such as immunofluorescence and Western blotting are described in section 8.4. Virus-infected cells can either be analysed directly by polyacrylamide gel electrophoresis (PAGE) and staining with Coomassie blue (for

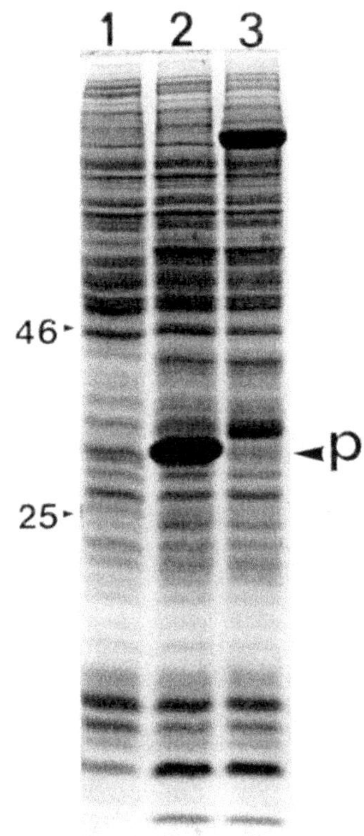

Figure 8.4 Analysis of proteins in insect cells using Coomassie blue staining. *S. frugiperda* cells were infected with a recombinant virus containing the *E. coli* β-galactosidase gene (AcMNPV.lacZ, Possee and Howard, 1987) or wild-type AcMNPV or were mock-infected with TC100 medium. The cells were harvested at 30 h p.i., and the proteins fractionated in a 12% PAG which was stained with Coomassie brilliant blue. Lane 1, uninfected Sf cells; lane 2, AcMNPV-infected Sf cells; lane 3, AcMNPV.lacZ-infected Sf cells. The proteins corresponding to β-galactosidase (β) and polyhedrin (P) are indicated.

an example, see Figure 8.4), or the synthesized proteins can be radiolabelled ([^{35}S]methionine is commonly used) prior to anlysis by PAGE (for an example, see Figure 8.5). If you cannot detect a foreign protein in a Coomassie-stained gel, try radiolabelling with an appropriate amino acid, as it may be that the levels of expression obtained are relatively low. Always include mock-infected and wild-type AcMNPV-infected controls on each gel; you will need to compare the profiles of these samples with those of the recombinant virus to be confident that the additional band of protein you see is not present in the control samples. Heavily glycosylated proteins can be difficult to detect, even when synthesized efficiently, as the protein can appear as a diffuse rather than tight band on gels. The presence of these proteins can be confirmed by immunological techniques (see section 8.4) or by inhibiting glycosylation with tunicamycin so that proteins with defined molecular weights are produced (see section 8.5).

The procedure below describes the preparation of radiolabelled or non-labelled protein extracts from Sf cells where foreign gene expression is driven by the polyhedrin or p10 promoters. We suggest that initially you examine protein synthesis at 24 and/or 48 hp.i., when using a polyhedrin or p10 promoter to drive gene expression. Each putative recombinant virus will need to be analysed. Once you have ascertained that one or more recombinant viruses are synthesizing the desired protein, and after amplification of a working stock of virus inoculum (see Chapters 6 and 7), you will probably need to examine the point of maximum synthesis of your foreign protein by undertaking a time-course experiment, where samples are pulse-labelled every 6 h or so for 72 h. Also, if you are using one of the earlier gene promoters to drive expression, remember to pulse-label or harvest samples at earlier time points e.g. about 16–18 h p.i. if using the basic protein promoter.

8.3.1 RADIOLABELLING PROTEINS IN VIRUS-INFECTED INSECT CELLS

When appropriate, cells may be left unlabelled and analysed in polyacrylamide gels which are subsequently stained with Coomassie brilliant blue.

Required:

1 × 35 mm dish of Sf cells for each virus to be analysed

2 × 35 mm dishes of Sf cells for controls (wild-type AcMNPV and mock-infected cells)

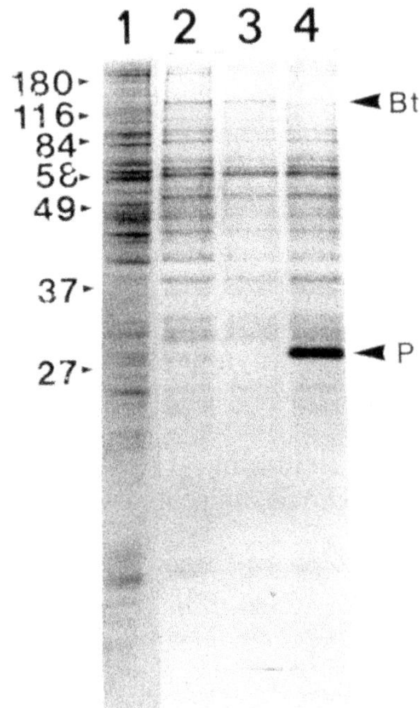

Figure 8.5 Analysis of proteins in insect cells using radiolabelling with [^{35}S]methionine. *S. frugiperda* cells were infected with a recombinant virus containing the *B. thuringiensis* delta endotoxin (Ac(PH$^-$)Bt; Merryweather *et al.* 1990) or wild-type AcMNPV or were mock-infected with TC100 medium. The cells were pulse-labelled with [^{35}S]methionine at 24 h p.i., prior to fractionation in a 12% denaturing PAG and autoradiography. Lane 1, uninfected Sf cells; lanes 2 and 3, Ac(PH$^-$)Bt-infected cells; lane 4, AcMNPV-infected cells. The proteins corresponding to delta endotoxin (Bt) and polyhedrin (P) are indicated. We thank Dr A. Clarke (née Merryweather) for supplying this figure.

TC100 growth medium
Starvation medium* (TC100 deficient in the appropriate amino acid e.g. methionine; see Chapter 5)
Radiolabelled amino acid (usually [^{35}S]methionine or [^{3}H]leucine)
PBS
TE buffer (10 mM Tris-HCl, pH 8.0, 0.1 mM EDTA)
* Some workers also add dialysed FCS to 2% (v/v)
Dissociation mix
50mM Tris-HCl, pH 6.9
25% (v/v) glycerol
10% (w/v) SDS
25% (v/v) 2-mercaptoethanol
0.25% (w/v) bromophenol blue

Method

1. Check that the cells have formed a sub-confluent monolayer. Remove the medium and inoculate the cells with 10 pfu virus/cell (recombinants and a wild-type AcMNPV control). Mock-infect 1 dish of cells with TC100 growth medium (100 μl). Leave the cells at room temperature for 1 h.
2. Remove the inoculum and discard into disinfectant. Overlay the cells carefully with 1.5 ml TC100 growth medium. Incubate in a moist sandwich box for the required period at 28°C.
   * The cells may be sampled at various times throughout infection (up to 72 h p.i.). When checking the efficiency of expression in a number of recombinant viruses for the first time, a good time point to use is 24 h p.i.
3. If the cells are not to be radiolabelled then proceed to step 8.
4. At the appropriate time (approximately 1.5 h before harvesting) carefully remove the medium from each dish. Drain the excess medium from the cell monolayer by standing one side of the culture dish on the lid for a few seconds.
   * After 18 h p.i. the infected cells start to detach from the culture dish. This can result in their loss when the culture medium is removed. For this reason we recommend that if an excessive number of cells are observed in the medium at this stage, they should be recovered by low-speed centrifugation. These may be resuspended in the starvation medium (5) and returned to the culture dish.
5. Add 0.5 ml starvation medium to each culture dish (see note in 4) and incubate the cells at 28°C for 30 min.
6. Add a further 0.5 ml starvation medium containing the appropriate radiolabelled amino acid (15–25 μCi for [^{35}S]-labelled, 1–5 μCi for [^{14}C]-labelled or 30 μCi for [^{3}H]-labelled amino acids).
7. Incubate the cells at 28°C for 1–2 h.
8. Scrape the cells from the culture dish into the labelling medium (or TC100 medium if cells are not radiolabelled) with the aid of a rubber policeman. Remove the cell suspension to a 1.5 ml microfuge tube and pellet the cells at low speed for 1–2 min.
9. Remove the labelling medium and discard safely (or TC100 medium if the cells are not radiolabelled). Resuspend the cells in 0.5 ml cold PBS and repellet.
10. Resuspend the cells in 100 μl TE buffer, add 25 μl dissociation mix and heat for 5 min at 100°C.
11. Analyse immediately using a denaturing polyacrylamide gel (see section 8.3.2) or store at −20°C until required.

8.3.2 POLYACRYLAMIDE GEL ELECTROPHORESIS OF INFECTED CELL EXTRACTS

There are numerous recipes available for the use of denaturing polyacrylamide gels to analyse proteins. The user of the baculovirus system will undoubtedly have considerable experience of these methods prior to attempting the expression of foreign proteins. For completeness, however, we will describe here a protocol for polyacrylamide gel electrophoresis, which we routinely use in our own laboratories for analysing recombinant protein production in insect cells. The method is intended for the larger type of polyacrylamide gel (20 × 20 cm) but can easily be adapted for the many excellent 'mini' protein gel systems now available. These latter methods have the advantage of producing more rapid results. This procedure is also a necessary pre-requisite for Western blot analysis of infected cell proteins.

Required:
- Gel buffer: 1 M Tris-HCl, pH 8.6
- 10% w/v SDS
- 10% Ammonium persulphate (APS, freshly prepared)
- *NNN'N'*-tetramethylethylenediamine (TEMED)
- Acrylamide/bis-acrylamide (30%/0.8%)
- Stacker buffer:
  - 0.5 M Tris-HCl, pH 6.9, 0.4% w/v SDS
  - Electrophoresis buffer (1×):
  - 185 mM glycine, 40 mM Tris, 0.1% w/v SDS
- Destain mix:
  - 5% glacial acetic acid, 50% methanol, 45% water
- Stain mix:
  - 0.2% w/v Coomassie brilliant blue dissolved in destain mix
  - Water-saturated butanol (or just water)
  - Gel plates, comb, spacers, etc. to prepare a 20 × 20 cm PAG
  - Protein samples from section 8.3.1
  - Protein size markers (e.g. purchase from BioRad, Amersham International)

Method:
1. Assemble clean glass plates to produce the mould for the gel.
2. For a 12% polyacrylamide gel mix the following (volumes will depend on type of gel apparatus used):

Acrylamide/bis-acrylamide	20 ml
Gel buffer	18.75 ml
Water	10.25 ml

10% w/v SDS	0.5 ml
10% w/v APS	0.5 ml
TEMED*	0.02 ml

* Add just before pouring the gel.

3. Pour the gel mix into the assembled glass plates to a point 1 cm below the expected position of the bottom of the comb, add a layer of water-saturated butanol (or just water) to the top of the mix to ensure even polymerization. Allow the gel to set for 30 min.
4. Remove the butanol layer and wash the top of the gel with water; remove surplus liquid with filter paper.
5. Insert the gel comb, leave at an angle to avoid air-bubble formation when adding stacking gel layer.
6. Mix stacker gel:

Acrylamide/bis-acrylamide	1.5 ml
Stacker buffer	2.5 ml
Water	6 ml
10% w/v APS	0.03 ml
TEMED*	0.03 ml

* Add just before use.

7. Add the stacker gel to the top of the resolving gel and check that there are no air-bubbles trapped under the comb. Adjust the comb to be parallel with the resolving gel. Allow the gel to set for 30 min.
8. Remove the comb and wash the sample wells with electrophoresis running buffer.
9. Assemble in the gel tank and add electrophoresis buffer to the lower and upper reservoirs.
10. Prepare protein samples as described in section 8.3.1. For a 20 $\times$ 20 cm polyacrylamide gel we usually analyse the equivalent of 2.5 $\times$ 10^5 cells per track. This may be adjusted according to the size of the gel. Include molecular weight size markers and control non-infected and wild type AcMNPV-infected tracks.
11. Run the gel overnight at about 30 V (for a 20 $\times$ 20 cm gel), until the bromophenol blue dye-front reaches the bottom of the gel.
12. Remove gel from plates and fix in destain mix for 1 h (radiolabelled samples). Alternatively, stain with Coomassie blue for 1 h immediately and then destain until protein bands are well defined. Agitate gel throughout these stages to ensure even staining/destaining.

    * When using [^{14}C]- or [^{3}H]-labelled amino acids, it will be necessary to process the gel for fluorography prior to drying and exposure to X-ray film. We have found that ENHANCE (Dupont) or Amplify (Amersham International) work well. With

[^{35}S]-labelled amino acids fluorography is not essential, but it does considerably reduce the exposure time.

13. Photograph the gel if required.
14. Dry the gel using a heated, vacuum apparatus and expose to X-ray film.

8.4 ANALYSIS OF RECOMBINANT PROTEIN SYNTHESIS IN INSECT CELLS USING IMMUNOLOGICAL TECHNIQUES

In circumstances where the concentration of the recombinant protein in infected cells is very low, it may be necessary to use more sensitive procedures to confirm expression. The methods described in this section all utilize antibody specific for the recombinant protein to detect expression using immunofluorescence, Western blotting or immunoprecipitation. The latter two methods are particularly useful to confirm the identity of a novel band(s) in a protein gel as the desired protein product, even when the level of expression is high enough to facilitate detection using polyacrylamide gel staining techniques or analysis of radiolabelled, infected cell extracts. An example of an immunoprecipitation experiment is shown in Figure 8.6. More detailed protocols for the following procedures can be found in the excellent laboratory manual, Antibodies (Harlow and Lane, 1988).

8.4.1 IMMUNOFLUORESCENCE

Immunofluorescence provides an extremely quick and reliable technique for monitoring recombinant protein expression in infected insect cells. The method may be varied to identify those proteins which are located on the outer surface of the cell or to locate cytoplasmic and nuclear proteins. A disadvantage of the method is that it is difficult to quantitate expression levels; analysis of protein levels using polyacrylamide gel electrophoresis is more appropriate for this purpose (see section 8.3).

Required:

1 × 60 mm dish of Sf cells (3 × 10^6 cells) for each virus to be analysed
TC100 growth medium
PBS
Specific antibody (polyclonal or monoclonal)
Second antibody conjugated to a fluorescent probe, e.g. tetra methylrhodamine or FITC
Multiwell slides (coated with 1% poly-L-lysine, Sigma)

Acetone
Methanol
UV microscope

Method:

1. Check that the cells have formed a sub-confluent monolayer. Remove the medium and inoculate cells with 10 pfu/virus/cell (recombinants and a wild-type AcMNPV control) (see note below). Mock-infect 1 dish with TC100 medium (100 μl). Leave the cells at room temperature for 1 h.
   * The moi may be varied between 0.01–5 pfu/cell. Some workers prefer to use a low moi to ensure that some cells in the sample remain uninfected and thus provide an internal, unlabelled control.
   * Prepare sufficient cells to analyse samples at 24 and 48 h p.i.
2. Remove the inoculum and discard into disinfectant. Overlay the cells carefully with 3 ml TC100 medium. Incubate the plates in a moist sandwich box at 28°C for the required time.
3. Harvest cells and pellet using a low-speed centrifuge. Wash once with PBS, repellet and resuspend in 100 μl PBS.
4. Add 5 μl of each sample to duplicate wells in multi-well slides (it will probably help if the slides have been coated with poly-L-lysine). Allow to air dry.
5. For confirmation that proteins expected to be targeted to the cell surface are located within the plasma membrane, do not fix the cells; proceed to step 6. For intracellular proteins, fix the cells in acetone (pre-chilled to −70°C) for 10 min or methanol (pre-chilled to −20°C) for 20 min. Air dry the slides.
6. Rehydrate the cells in PBS for 5 min at room temperature and then carefully blot the excess buffer from the slide (take care not to remove cells).
7. Add sufficient primary antibody (appropriately diluted in PBS) to cover the wells of the slide.
8. Incubate at 37°C for 30 min in a moist atmosphere.
9. Rinse the slide five times with PBS and then soak in PBS for 5 min. Rinse a further five times in PBS.
10. Repeat steps 7, 8 and 9 with the second antibody that is conjugated to the fluorescent probe.
11. Dry the slides and examine with a UV microscope using the appropriate wavelength and filter depending on the fluorescent probe used.
12. Photograph the cells. Proteins that are membrane targeted should give rise to a halo of fluorescence around the cell (non-fixed cells) and proteins that are localized intracellularly should give rise to the whole cell fluorescing.

8.4.2 WESTERN BLOT ANALYSIS OF VIRUS-INFECTED CELL PROTEINS

We describe here two methods whereby proteins may be transferred to a membrane using electrophoresis. The first, a 'semi-dry' blotting technique, requires only one hour to achieve the transfer of proteins to a nitrocellulose membrane. The second, the 'wet' blotting technique, normally requires between 6 and 12 h, but is considered by some to achieve a more efficient transfer than the semi-dry method.

Required:

Recombinant virus-, AcMNPV-infected and uninfected cell extracts: these should be prepared as described in section 8.3.1
Semi-dry blotting apparatus (e.g. Sartorius) or,
Wet blotting apparatus (e.g. Bio-Rad)
Whatman 3MM chromatography paper
Nitrocellulose membrane (0.45 μM pore size)
Transfer buffer (25 mM Tris-HCl, pH 8.3, 192 mM glycine, 20% w/v methanol)

Method:

1. Prepare infected and uninfected cell extracts as described in section 8.3.1 and fractionate in a polyacrylamide gel as described in section 8.3.2. Do not fix the gel; proceed with semi-dry or wet blotting as outlined below.

A. Semi-dry blotting

2. Wet the graphite electrode plates and place six sheets of Whatman 3MM paper (the same size as the polyacrylamide gel), pre-soaked in transfer buffer, onto the lower (cathode) plate. Take care to remove trapped air from between the sheets.
3. Place the polyacrylamide gel onto the Whatman 3MM sheets and then cover with the nitrocellulose membrane (pre-soaked in transfer buffer). Complete the sandwich with six more sheets of Whatman 3MM (pre-soaked in transfer buffer).
4. Assemble the apparatus by placing the top (anode) electrode plate on the sandwich. Ensure that the nitrocellulose membrane is adjacent to the anode.
5. Perform electrophoresis at 0.8 mA/cm^2 gel surface for 1 h.
6. After completion of electrophoresis, dismantle the apparatus and proceed to section C.

B. Wet blot (Bio-Rad apparatus)

1. Wet a Scotch-Brite pad with transfer buffer and displace air bubbles. Cut a piece of Whatman 3MM paper to the same size as the pad, soak with transfer buffer and place it on the pad.

2. Position the polyacrylamide gel on the Whatman 3MM paper and then place a sheet of nitrocellulose membrane (cut to match the size of the gel and pre-soaked in transfer buffer) onto the gel.
3. Cover the nitrocellulose with another sheet of Whatman 3MM, pre-soaked in transfer buffer, and complete the sandwich with another Scotch-Brite pad. If the Bio-Rad apparatus is being used then place the gel/nitrocellulose assembly in the hinged plastic holder and slide it into the electrophoresis tank.
   * It is important at every stage of the assembly process described above to remove air trapped between the respective components. If this precaution is not exercised then even transfer of proteins from the gel to the membrane will not occur.
4. Fill the electrophoresis tank with transfer buffer (pre-chilled to 4°C) and begin the protein transfer at 4°C. Use 30 V, 0.1 A for overnight runs and 60 V, 0.22 A for runs of less than 6 h.
5. After the completion of transfer, dismantle the apparatus and proceed to the next section.

C. Detection of proteins on nitrocellulose membranes
Required:
- Blocking solution (5% (w/v) powdered milk, 0.05% Tween 20 in PBS)
- Antibody specific for the recombinant protein
- Anti-antibody conjugated with alkaline phosphatase
- PBST (0.05% (v/v) Tween 20 in PBS)
- Staining buffer (0.1 M Tris-HCl, pH 9.5, 0.1 M NaCl, 50 mM $MgCl_2$)
- NBT and BCIP (e.g. from Gibco-BRL)
- TE buffer (10 mM Tris-HCl, pH 8.0, 1 mM ETDA)

Method:
1. Wash the nitrocellulose membrane with a small volume of blocking solution for 1 h at room temperature.
2. Dilute the antibody specific for the recombinant protein in blocking solution (e.g. 1/500–1/10,000) to give the optimum concentration (determined empirically) and wash the filter with this solution for 1 h at room temperature. Ensure that the side of the filter bearing the protein is uppermost and that the filter does not dry out or stick to the bottom of the container.
3. Remove the antibody solution and wash the filter four times with PBST (15 min for each wash).
4. Dilute the anti-antibody in blocking solution as required and wash the filter at room temperature for 1 h.
5. Repeat step 3.
6. To develop with alkaline phosphatase, equilibrate the filter with staining buffer and then stain in the dark with NBT (50 μl) and

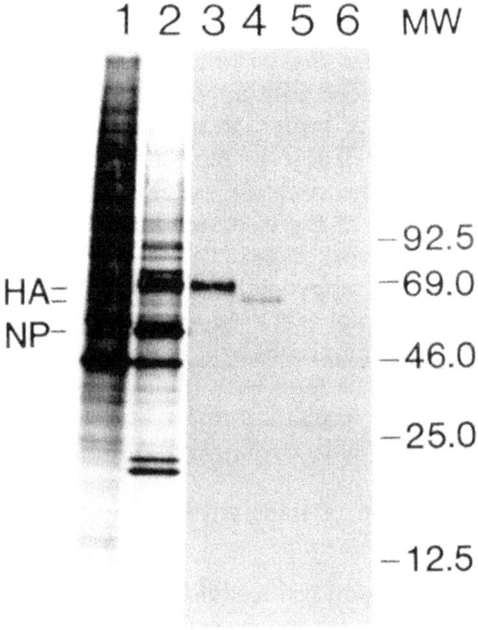

Figure 8.6 SDS-polyacrylamide gel of influenza virus haemagglutinin (HA) precipitated by anti-HA monoclonal antibody from [^{35}S]methionine-labelled insect cell extracts. Lanes 1 and 2 show total cell extracts from mock-infected and influenza virus-infected BHK cells (controls), respectively. Immunoprecipitates: lane 3, influenza virus-infected BHK cells; lane 4, recombinant baculovirus, AcRP5-PR8.HA (Possee, 1986) infected Sf cells; lane 5, wild-type AcMNPV-infected Sf cells; lane 6, mock-infected Sf cells. The positions of molecular weight size markers are shown in kDa and the positions of the two sizes of HA and the influenza virus nucleoprotein (NP) are indicated. Reproduced from Possee (1986) with permission.

BCIP (60 μl) in 20 ml staining buffer. Stop the reaction when desired by rinsing in TE buffer.

7. Photograph the membrane.

8.4.3 IMMUNOPRECIPITATION OF VIRUS-INFECTED CELL PROTEINS

This represents an alternative approach to Western blotting for identifying the recombinant protein in virus-infected cell extracts. Infected cells are lysed with detergent and incubated with an antibody specific for the recombinant protein. By altering the conditions used to lyse the infected cells, those proteins in the whole cell, the nucleus or cytoplasmic fraction may be identified. The antigen-antibody complexes are then precipitated with protein A bound to an insoluble matrix and analysed in standard polyacrylamide gels (see Figure 8.6). This method requires significantly more manipulations than Western blotting, but has the advantage of a greater sensitivity since the radiolabelled recombinant protein is concentrated throughout the procedure. It is necessary to pulse-label the virus-infected cells for a considerably longer period to ensure that sufficient material is available for detection of the recombinant protein. This increases the proportion of other virus and cell-specific proteins which are radiolabelled, but the very nature of the immunoprecipitation technique ensures that most of these are removed prior to electrophoresis.

Required:

Radiolabelling of insect cells:

- 1 × 60 mm dish of Sf cells (3 × 10^6 cells) for each virus to be analysed
- 2 × 60 mm dishes of Sf cells (3 × 10^6) for controls (AcMNPV and mock-infected cells)
- TC100 growth medium
- Starvation and labelling medium: TC100 minus bactotryptose broth and the amino acid to be used for radiolabelling (see Chapter 5); dialysed FCS may be added to 2% (v/v)

Cell lysis:

- Non-ionic detergent lysis buffer (NTEP; 50 mM Tris-HCl, pH 7.9, 150 mM NaCl, 5 mM EDTA, 0.5% v/v Nonidet P40)
- Ionic-detergent lysis buffer (NTES; 50 mM Tris-HCl, pH 7.9, 150 mM NaCl, 5 mM EDTA, 0.5% v/v Nonidet P40, 0.1% w/v SDS)

Method:

A. Radiolabelling and cell fractionation

1. Check that the cells have formed a sub-confluent monolayer. Remove the medium and inoculate cells with 10 pfu virus/cell (recombinants or AcMNPV). Mock-infect 1 dish with TC100 growth medium (100 μl). Leave the cells at room temperature for 1 h.
2. Remove inoculum and discard into disinfectant. Overlay the cells carefully with 1.3 ml growth medium. Incubate the plates in a moist sandwich box for for 14 h.
3. Replace growth medium with 1 ml labelling medium and incubate at 28°C for 1 h.
4. Add 1 ml growth medium containing 50 μCi [^{35}S]methionine, 100 μCi [^{3}H]leucine or 10 μCi [^{14}C]amino acid. Incubate at 28°C for 8 h.

* The above procedure ensures maximum radiolabelling of the virus proteins induced in the latter stages of the infection, but avoids contaminating the preparation with too many cellular and virus proteins produced in the early phases of the infection cycle.

5. Harvest the cells and pellet using a low-speed microfuge. Wash once with 1 ml PBS and then proceed to either of the cell fractionation methods.

B. Cell lysis

Two methods are suggested here. In the first, a mild detergent treatment is used to lyse the cells but leave the nuclei intact. The second approach uses an ionic detergent completely to disrupt all organelles within the cell.

Lysis method 1: Extraction of proteins associated with the cytoplasm or nucleus

1. Resuspend the radiolabelled cells in 1 ml NTEP and stand on ice for 5 min.
2. Freeze–thaw the cells three times using dry-ice and ethanol in combination with a 37°C water bath.
3. Pellet the nuclei using a microfuge (5 min at high speed) and remove the supernatant fraction to a fresh tube. If the proteins within the nuclear fraction are to be analysed, resuspend the nuclei in 1 ml NTEP and stand on ice for 5 min. Pellet the nuclei as before and discard the supernatant fraction. The nuclei may be treated with the ionic detergent buffer to liberate nuclear proteins (see section 8.5).

Lysis method 2: Extraction of total cell or nuclear proteins.

1. Resuspend the radiolabelled cells, or isolated nuclei in 1 ml NTES, and stand on ice for 5 min.
2. Pellet the lysate using a microfuge (5 min at high speed) and remove the supernatant fraction to a fresh tube.

C. Formation of antigen–antibody complexes

Required:

Antigen extract (see A or B above)
Protein A-insoluble, lyophilized cell powder (Sigma, P 9151; prepared as a 10% (w/v) suspension in NTEP or NTES as appropriate)
Antibody specific for the recombinant protein
Pre-immune serum
5× protein gel dissociation mix (see section 8.3.2)

Method:

1. Divide the antigen extract into two equal volumes and add 50 μl

of the protein A suspension to each. Incubate with shaking for 30 min at room temperature and then pellet the bacterial suspension at 12 000 *g* for 5 min using a microfuge.

2. Transfer the supernatant fraction to a fresh tube and repeat step 1.
3. Add the specific antibody to one tube, and pre-immune serum to the second. Incubate at room temperature for 2 h with shaking.
   * The concentration of antibody to be used will have to determined empirically.
   * If the Fc portion of the specific antibody which recognizes your recombinant protein binds protein A very poorly (e.g. chicken, goat and some rat and sheep antisera), then it may be necessary to use an anti-immunoglobulin antibody after this step to provide a good attachment site.
4. Add 50 μl of the protein A suspension to each tube and continue to shake at room temperature for a further 2 h.
   * Protein A bound to Sepharose beads may be used in place of the crude bacterial preparation; this reportedly gives better backgrounds but is expensive. We routinely use the cheaper method with satisfactory results.
5. Pellet the bacterial suspension at 12 000 **g** for 5 min in a microfuge.
6. Wash the pellets three times with 500 μl NTEP or NTES.
7. Resuspend the final pellets in 100 μl TE, add 25 μl 5× protein dissociation mix and analyse using gel electrophoresis (section 8.3.2).
   * High backgrounds visible after autoradiography of the protein gel may be removed by pretreating the infected cell lysate with a pre-immune serum and protein A before using the immune serum. This adds considerably to the time required for the technique but may be worthwhile if poor results are obtained with the standard method.

8.5 ANALYSIS OF POST-TRANSLATIONAL PROCESSING EVENTS IN INSECT CELLS

The ability of the baculovirus expression system to perform the appropriate post-translational modifications of recombinant proteins was briefly reviewed in Chapter 3. In this section we describe some of the most commonly used protocols for analysing these events.

8.5.1 GLYCOSYLATION

This appears to be the one area where insect cells differ significantly from mammalian cells. In consequence it provokes the most concern from users of the baculovirus system. Two approaches may be considered to investigate glycosylation. In the first, the virus-infected cell proteins are radiolabelled with a sugar precursor (usually [^{14}C]mannose) and analysed using polyacrylamide gels and autoradiography; essentially the method is as described for the use of radiolabelled amino acids (section 8.3). In the second, inhibitors of the glycosylation process in insect cells are employed. These inhibitors are particularly useful when comparing the sizes of the recombinant protein produced in insect cells and the normal (probably vertebrate) host cell. When the proteins are produced without post-translational glycosylation their size should be identical, irrespective of the host cell. Furthermore, this method also serves to resolve problems caused by variable glycosylation of proteins in insect cells. This manifests itself as multiple bands on protein gels. Treatment of the infected cells with an inhibitor of glycosylation, such as tunicamycin, should reduce these bands to a single species.

A. Radiolabelling of glycoproteins in insect cells
Required:

As described in section 8.3.1, plus:
TC100 with 10% the normal concentration of glucose (see Chapter 5) and without bactotryptone broth
[^{14}C]Mannose (e.g. from Amersham International)

Method:
The method used is essentially the same as that described in section 8.3.1 when radiolabelling cells with [^{35}S]methionine, etc. The cells should be radiolabelled with 10 μCi [^{14}C]mannose for 1 h prior to analysis in polyacrylamide gels (section 8.3.2). It is recommended that the gel is processed for fluorography.

B. Inhibitors of N-linked glycosylation
N-linked glycosylation occurs in the endoplasmic reticulum where oligosaccharides are attached to asparagine residues. N-linked glycosylation is sensitive to inhibition by tunicamycin in baculovirus-infected cells (Kelly and Lescott, 1983).

Required:

As described in section 8.3.1, plus
Tunicamycin (Sigma) (1 mg/ml stock)

Method:

1. Prepare virus-infected and control cells as described in 8.3.1

and incubate at 28°C until 6 h prior to the time of radiolabelling.

2. Replace the growth medium with fresh medium containing tunicamycin at a concentration of 1 μg/ml. Continue the incubation at 28°C for 6 h.
3. Fifteen minutes before adding the radioisotope ([^{35}S]methionine, etc.), replace the growth medium with 0.5 ml starvation medium plus tunicamycin. Incubate at 28°C for 15 min.
4. Add a further 0.5 ml starvation medium containing the radioisotope and tunicamycin.
5. Incubate the cells at 28°C for 1 h and harvest the cells as described in section 8.3.1.
6. Analyse the cell extracts using the method for polyacrylamide gel electrophoresis described in section 8.3.2.

8.5.2 PHOSPHORYLATION

Required:

As described in section 8.3.1, plus
Phosphate labelling medium: TC100 minus phosphate plus 2% dialysed FCS (see Chapter 5)
Carrier-free [^{32}P]

Method:

1. Infect cells as described in section 8.3.1, but after the 1 h adsorption period overlay the plates with phosphate labelling medium. Incubate at 28°C until the time of radiolabelling.
2. Replace medium with fresh phosphate labelling medium containing 10 μCi [^{32}P]. Incubate for 1 h at 28°C.
3. Harvest cells and analyse proteins as described previously (section 8.3.1 and section 8.3.2).

8.5.3. PALMITYLATION AND MYRISTYLATION

Required:

As described in section 8.3.1, plus:
Dimethylsulphoxide (DMSO)
TC100/0.25% bovine serum albumin (BSA; fatty-acid free)
[^{3}H] palmitic acid (supplied in toluene)*
[^{3}H] myristic acid (supplied in toluene)*
* Just prior to use, dry down the required radiolabelled fatty acid and redissolve in DMSO plus 0.25% BSA. Dilute the radiolabelled fatty acid in TC100/0.25% BSA to give a final concentration of 1% DMSO.

Method:

1. Infect cells as described in section 8.3.1 and incubate at 28°C until the time of radiolabelling.
2. Replace the growth medium with TC100/0.25% BSA and incubate at 28°C for 1 h.
3. Remove the TC100/0.25% BSA and add 1 ml of the same medium containing the 50–100 μl radiolabelled fatty acid. Incubate at 28°C for 1–4 h.
4. Harvest the cells and analyse the proteins as described previously (section 8.3.1 and 8.3.2)

8.6 ANALYSIS OF TRANSCRIPTION IN RECOMBINANT VIRUS-INFECTED CELLS

Analysing the amount of mRNA produced by the recombinant virus in insect cells is a useful exercise if the yields of foreign protein, as assessed by protein gels, are low. Although the reasons for such poor results are unclear, it is generally concluded at present that the nature of the recombinant product itself is the major determinant of the level of expression attained in insect cells. However, it is as well to confirm that there is no problem with the transcriptional machinery of the virus; a random mutation in the polyhedrin promoter or in other regulatory elements might reduce mRNA levels. The simplest way of checking transcription is to extract total RNA from infected cells and perform Northern hybridization analysis. This method will only measure steady-state RNA levels in infected cells; any problems with mRNA stability would have to addressed using more sophisticated approaches which are beyond the scope of this book.

We have located these protocols after the sections which addressed the analysis of the recombinant virus genome and the levels of protein synthesis in infected cells. Although out of sequence, it was considered appropriate to position this section at the end of the chapter because few workers will analyse transcription events before testing protein expression levels.

In common with some of the other methods we have described, there are a multitude of different protocols for extracting RNA from cells. The method presented here has been used routinely in our laboratories with consistent success. Essentially it is based on disrupting cells with guanidinium isothiocyanate and then removing protein and DNA in the presence of hot phenol in acidic conditions. A disadvantage is the use of hot phenol, which is a hazardous reagent and must be treated with respect (work in a fume hood, wear gloves, labcoat and protective glasses, etc.). The method does, however,

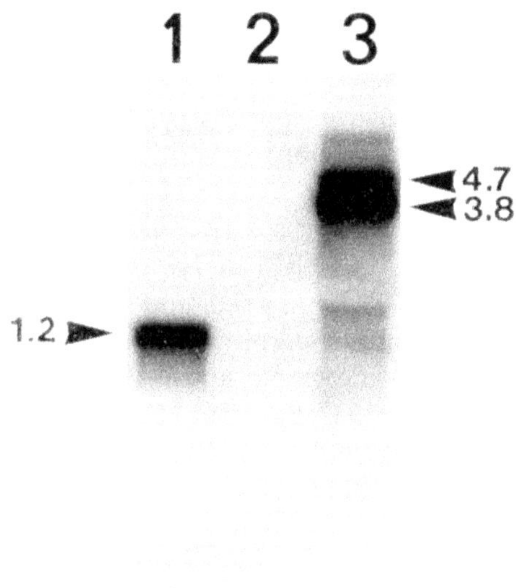

Figure 8.7 Northern blot hybridization analysis of insect cell RNA. *S. frugiperda* cells were infected with AcMNPV, a recombinant virus encoding *E. coli* β-galactosidase (AcMNPV.lacZ; Possee and Howard, 1987) or were mock-infected with TC100 medium. The cells were harvested at 30 h p.i. and total RNA extracted using the method described in section 8.6. A sample of each RNA was glyoxylated prior to fractionation in a 1% agarose gel and transferred to a Genescreen nylon membrane. The immobilized RNA was then hybridized with radiolabelled probes specific for the polyhedrin gene (lane 1, AcMNPV; lane 2, mock-infected Sf cells) or the β-galactosidase gene (lane 3, AcMNPV.lacZ). The membrane was washed in SSC and exposed to an X-ray film. The sizes of the RNA species detected are indicated in kb.

largely remove the problem of ribonuclease contamination since these ubiquitous enzymes are inactivated by the harsh denaturing conditions used. Furthermore, the volumes used in this protocol are minimal. The method described below was adapted from Maniatis *et al.* (1982) and Sambrook *et al.* (1989). An example of a Northern blot, using the protocols outlined in this chapter, is shown in Figure 8.7.

8.6.1 EXTRACTION OF RNA FROM INSECT CELLS

Required:
1×60 mm dish of Sf cells (3×10^6 cells) for each virus to be analysed

2 × 60 mm dishes of Sf cells (3 × 10^6 cells) for controls (wild-type AcMNPV and mock-infected cells)
TC100 growth medium
Virus stocks for inoculation (preferably of known titre, see Chapter 6 for more detailed instructions for setting up and inoculating cells)
PBS
Guanidinium isothiocyanate solution. Dissolve 100 g guanidinium isothiocyanate (e.g. Gibco-BRL) in 100 ml deionized, sterile water, 10.6 ml of 1 M Tris-HCl, pH 7.6 and 10.6 ml of 0.2 M EDTA. The mixture may require warming to 37°C. Add 21.2 ml of 20% (w/v) sodium lauryl sarkosinate and 2.1 ml of 2-mercaptoethanol. Adjust the volume to 212 ml with deionized, sterile water. Filtering the solution through a disposable Nalgene filter is sometimes recommended; we have not found this to be necessary. Store at 4°C.
Sodium acetate/EDTA solution (0.1 M sodium acetate, pH 5.2, 1 mM EDTA)
3 M sodium acetate, pH 5.2
Phenol* equilibrated in water
Phenol*/chloroform (50:50) equilibrated in water
*The phenol should be redistilled or purchased as an ultra-pure RNA quality reagent.
75 and 100% ethanol
Deionized sterile water

Precautions:
* Use disposable plastic reagents when possible to avoid problems with ribonucleases. Where glassware has to be used with organic solvents it should be heat-treated at 180–200°C overnight prior to use.
* Wear gloves at all times when working with RNA to avoid ribonuclease contamination from the fingers.
* Have a set of reagents set aside for work with RNA. Repeated use of 'clean' spatulas when dispensing chemicals is a recipe for disaster.

Method:

1. Check that the cells have formed a sub-confluent monolayer. Remove the medium and inoculate cells with 10 pfu virus/cell (recombinants and a wild-type control). Mock-infect one dish with TC100 medium (500 μl). Leave the cells at room temperature for 1 h.
2. Remove inoculum and discard into disinfectant. Overlay cells carefully with 3 ml growth medium. Incubate the plates in a moist sandwich box for 36 h at 28°C.
3. Harvest the cells at 36 h p.i. and pellet using a low speed on

a bench-top centrifuge. Remove the medium and discard into disinfectant.
4. Wash the cells in PBS and re-pellet.
5. Resuspend the cells in 250 μl guanidinium isothiocyanate mix and heat the mixture to 60°C.
6. Add 250 μl phenol/water and 250 μl sodium acetate/EDTA. Mix vigorously by vortexing and then stand in a 60°C water bath or heating block. Incubate for 10 min at 60°C; continue to mix well about every 30 seconds.
7. Add 250 μl chloroform, mix briefly and cool to room temperature. Separate the phases by centrifugation at 12 000 *g* in a microfuge for 5 min.
8. Remove the aqueous phase to a fresh tube and add 500 μl phenol/chloroform; mix for 2 min then separate the phases as in 6.
9. Remove the aqueous phase to a fresh tube and add 500 μl chloroform; mix briefly. Separate the phases by briefly spinning in a microfuge, as in 6.
10. Remove the aqueous phase to a fresh tube and add 50 μl 3 M sodium acetate, followed by two volumes of ethanol. Chill at −20°C for at least 5 h or overnight.
11. Pellet the RNA at 12000 g in a microfuge for 10 min. Wash the pellet twice with 75% ethanol.
12. Dry the pellet using a vacuum desiccator or by air drying.
13. Resuspend the RNA in 20 μl water and store at −70°C.
    * The RNA prepared by this method is essentially free of DNA contamination.

8.6.2 ANALYSIS OF RNA USING NORTHERN BLOT HYBRIDIZATION

The method recommended is essentially that described by McMaster and Carmichael (1977) and Thomas (1980). It utilizes glyoxal and DMSO to denature RNA prior to transfer to a nylon membrane. Nitrocellulose, the traditional choice for DNA/RNA transfers, may be used but is significantly less robust than the nylon equivilants.

Required:

Glyoxal. Glyoxal is purchased as a 40% aqueous stock (Sigma) and deionized prior to use. As this step is essential to the success of the technique, it will be described in full

* Plug the narrow end of a 10 ml disposable syringe with heat-treated glass wool. Fill the syringe with Amberlite MB-3 (Sigma) to the 8 ml point. Add about 5 ml crude glyoxal to the syringe and

collect as it elutes from the column. Pass the glyoxal repeatedly over the resin and monitor the pH using pH papers. Initially the pH will be less than 3, but should gradually rise to about 6. When the pH stabilizes, begin collecting the pure glyoxal in small (0.5 ml), autoclaved microfuge tubes. Fill each one to the top and leave the minimum amount of air above the liquid. It is important to collect the glyoxal as it elutes from the resin; do not drain the syringe into a larger tube and then divide the glyoxal into aliquots. Store the glyoxal at −20°C. Use one aliquot at a time and discard after a single use

* Do not attempt to purify glyoxal by stirring with resin in a beaker or other vessel; this does not seem to work as efficiently as the syringe method

* The quality of the glyoxal is probably the single major determinant of the successful outcome of an RNA analysis using this method

Dimethyl sulphoxide: Purchase analytical grade reagent, this may be used without further treatment

Phosphate buffers: Na_2HPO_4 and NaH_2PO_4 (usually purchased as hydrated stocks). Prepare separate, 0.2 M stocks in deionized, sterile water and mix to obtain a 0.2 M buffer of pH 6.5. This may then be diluted according to requirements

1% (w/v) agarose: prepare in deionized, sterile water and autoclave

* The recommended concentration is only a rough guide; the size of your recombinant mRNA will determine your own choice

5× RNA sample buffer: 25% glycerol, 10 mM phosphate buffer (pH 6.5), trace of bromophenol blue

Hybridization membranes; Genescreen (Dupont), Hybond N (Amersham International),

20× SSC: 3 M NaCl, 0.3 M tri-sodium citrate

Hybridization buffer: 50% deionized formamide (pH to 7.0 with Amberlite MB-3,Sigma), 5× SSC, 50 mM phosphate buffer, pH 6.5, 0.02% polyvinylpyrrolidone, 0.02% ficoll, 0.02% bovine serum albumin. Prepare as a stock solution and store at 4°C. Just prior to use add 10 μg/ml sonicated salmon testes DNA

[^{32}P]-labelled DNA probe (denature for 3 min in a boiling water bath, then rapidly cool in dry ice/ethanol bath, take care not to freeze the mixture)

Materials for agarose gel electrophoresis, including a pump to recirculate the running buffer

* Methods appropriate for the preparation of hybridization probes may be found in most of the currently available guides for molecular biology techniques, e.g. Sambrook *et al.* (1989).

Method:

1. Prepare an agarose gel in a suitable electrophoresis tank and soak in 10 mM phosphate buffer for at least 1 h.
2. Mix:

RNA (from section 8.6.1)	5 μl
Glyoxal	3.5 μl
DMSO	10 μl
Water	1.5 μl

3. Incubate at 50°C for 1 h.
4. Add 5 μl 5× RNA sample mix and load samples in a prepared 1% agarose gel.
   * Molecular weight standards should be included. Most conveniently, these may consist of λ DNA digested with both *Eco*RI and *Hin*dIII and radiolabelled with [^{32}]P by a 'fill-in' reaction with the Klenow fragment of *E. coli* DNA polymerase (see Sambrook *et al.*, 1989). These will be transfered to the hybridization membrane and readily visible after autoradiography.
5. Electrophorese samples until the bromophenol blue is 1 cm into the gel.
6. Begin recirculation of the electrophoresis buffer.
   * Buffer recirculation is important to prevent the pH rising above 7.0; glyoxal dissociates from RNA at pH 8.0 or higher.
7. The gel may be run overnight at 30 V (for 20 × 20 cm gel) or for about 1 h if it is a mini-gel.
8. Remove the gel from the tank and trim excess agarose. Do not stain the gel with ethidium bromide since glyoxalated RNA does not bind ethidium bromide efficiently; the stain also reduces the transfer of the RNA to the hybridization membrane.
9. Place gel on a transfer apparatus assembled as described in Figure 8.3.
   * The transfer buffer used will depend on the choice of hybridization membrane. For Genescreen, use 25 mM phosphate buffer, pH 6.5; for Hybond, use 20× SSC.
10. Transfer RNA to hybridization membrane (overnight for a large gel, 2–3 hours for a mini-gel)
11. Remove gel and membrane from transfer apparatus and mark positions of wells on the membrane. Discard gel.
12. Do not wash the hybridization membrane. Bake immediately at 80°C for 2 h.
    * It is necessary briefly to bake the membrane to dissociate the glyoxal from the RNA, so do not use UV cross-linking to bind the RNA onto the membrane.
13. Soak the membrane in sterile water for 15 min and then boil in 1 mM EDTA for 5 min.

* Use a microwave oven for this operation. This step is essential to remove the salt if 20× SSC was used as the transfer buffer. The boiling process also removes the need to perform a prehybridization step; background signals detected during autoradiography are considerably reduced.

14. Place the membane in a stout plastic bag and seal on three sides with a heat sealer, or use a commercial hybridization bottle.
15. Add the hybridization solution (see above) and seal the fourth side of the bag. Place in a second bag and seal.
    * Use approximately 20 ml solution for a 20 × 20 cm filter.
16. Incubate in a shaking water bath or hybridization oven at 42°C for 16 h.
17. Remove the membrane from the bag or bottle and wash:
    Four times with 2× SSC/0.1 % SDS at 65°C for 5 min.
    Twice with 0.1× SSC/0.1 % SDS at 50°C for 15 min.
    * The above washing procedure is brief, but has been found to remove most of the background signals from the membrane.
18. Seal the washed membrane in a plastic bag before it dries. If the membrane is maintained in a damp condition then the hybridization probe may be easily removed. Dry filters are more difficult to strip.
19. Expose the filter to X-ray film overnight and develop the autoradiograph.
    * The amount of RNA produced by AcMNPV-infected cells is considerable and exposure times of less than 6 h are sometimes sufficient to detect a strong band of RNA on a membrane.

Scaling up the production of recombinant protein in insect cells; laboratory bench level 9

9.1 INTRODUCTION

After characterizing the recombinant virus and confirming that it is synthesizing the expected product (see Chapter 8), the next step will almost certainly be to scale up the production of protein. It is at this point that our confidence in the available methods decreases. There is very little published information about this topic. Furthermore, most academic laboratories, our own included, do not need to produce very large quantities of recombinant protein. From informal discussions, we are aware that industry, and a few academic laboratories, are beginning to address the problem, but few results have reached the press to date. This is probably the one subject area which requires the most attention in the next few years.

One system that looks promising for the scale-up of insect cells is a cytoculter fermenter (1 l size) made by SGI Ltd (see Appendix A). Tests carried out in at least two laboratories have shown that Sf9 and Sf21 cells can be grown routinely to densities in excess of 6×10^6/ml, when using medium supplemented with 0.1% pluronic (Steve Reid, Department of Chemical Engineering, The University of Queensland, Australia and Professor D. H. L. Bishop, NERC Institute of Virology and Environmental Microbiology, Oxford, personal communication). The SGI fermenter consists of a borosilicate glass vessel with an impeller driven by a variable speed motor. Various entry ports include an air inlet and outlet, temperature probe, pH probe, O_2 probe, CO_2 probe, four inoculation, two sampling and a cooling tube. Experiments are in progress to determine the ability of this system to produce high-titre virus stocks and high levels of recombinant proteins.

In this chapter, the methods described are concerned mainly with the degree of scale-up required by most academic users of the baculovirus system. The equipment needed for these protocols is very cheap to acquire and is well within the budget of most research groups. In our experience, most laboratories, with a little imagination, can adapt their present stocks of glassware, stirrers, incubators, etc. to fulfil the protocols described below. We will also consider several other factors which we consider to be particularly important when trying to maximize protein production in large cultures of insect cells (e.g. cell density, multiplicity of infection, time of harvesting infected cells, etc.)

The numbers of infected cells required will be determined by the eventual use of the recombinant material. For instance, if it is to be used to derive an antiserum in mice or rabbits then very little protein may be required. In some cases, infected whole cell homogenates have been used for the immunization procedure, which greatly simplifies the experimental protocols. Conversely, if the protein is required for a vaccine trial in animals it may be necessary to isolate the product in a form which is free of contaminating insect cell proteins. The ultimate purity will be required for protein crystallographic studies. These experiments will also require milligram amounts of protein which will probably need to be produced using fermentation processes; these techniques are beyond the scope of the present book. The number of cells required for extracting the recombinant protein will also be dependent on the initial concentration of the product in infected cells and the efficiency of the extraction process.

In consequence, there is no standard procedure to adopt when designing your scale-up experiments. You will need to consider the various factors mentioned above (use of the protein, purity required, concentration and ease of extraction) before embarking on the most appropriate route. Essentially there are two options available for growing large quantities of insect cells: monolayer or suspension cell culture.

Suspension cultures of insect cells undoubtedly have the potential to yield considerably higher numbers of cells than monolayer cultures. When attempting to produce milligram to gram quantities of protein this will be the option of choice. The disadvantage of this method is that difficulties are sometimes encountered when attempting to establish a synchronous virus infection. This will be discussed below (section 9.5). For the newcomer to the baculovirus expression system the easiest option is to make use of monolayer cultures of insect cells. It is very easy to infect these cells synchronously.

9.2 LARGE-SCALE CULTURE OF INSECT CELLS

9.2.1 LARGE-SCALE CULTURE OF INSECT CELLS IN MONOLAYER CULTURES

The use of monolayer cultures to grow insect cells has already been described (see Chapter 5). Stoppered, plastic flasks are available which have surface areas of between 25–225 cm^2. An alternative is the use of cell culture quality Petri dishes. These are substantially cheaper than flasks and, in our opinion, easier to use; few problems have been encountered with contamination.

Another option is plastic roller bottles, which may have surface areas of 1750 cm^2. These require some facility for turning the bottles in a regulated manner. Furthermore, the roller equipment will almost certainly have to be mounted in a dedicated 28°C room; the number of roller bottles which could be accommodated in a standard laboratory incubator would present few advantages over plastic flasks or dishes. Weiss *et al.* (1981) described a semi-automated procedure for growing Sf cells using roller bottles, but this is likely to be beyond the capabilities of most laboratories. The price of the plastic roller bottles is also likely to be prohibitive. We have tried to use glass roller bottles, but the Sf cells adhere very poorly to the walls of these vessels. In conclusion we recommend the use of multiple plastic Petri dishes for the preparation of recombinant protein in the first instance.

Apart from cost considerations, the chief disadvantage of using large-scale monolayer culture is that substantial quantities of cells have to be made available for seeding the requisite number of dishes. For example, if each 150 cm^2 Petri dish is seeded with 3×10^7 cells, then 20 such dishes would require 6×10^8 cells. To produce this number of cells using standard monolayer techniques would be tiresome and wasteful of labour and materials. It is better to make use of the suspension culture technique that was described in Chapter 5 to produce the uninfected cell inoculum for seeding the monolayer cultures. Despite these disadvantages, for a small-scale production of a recombinant protein, monolayer cultures can be appropriate.

Required:

1–n × 500 ml suspension cultures seeded with 10^5 Sf cells/ml (see Chapter 5 for details of how to set up this culture)
150 cm^2 Petri dishes
Insect cell medium (e.g. TC100/FCS)

Method:

1. Grow the suspension culture at 28°C until the cell density reaches

$1–2 \times 10^6$ cells/ml. This density represents a point just before the cells enter the stationary phase.

2. Seed each 150 cm^2 Petri dish with 3×10^7 Sf cells from the suspension culture and adjust volume to 30 ml with growth medium, if necessary.
3. Incubate at 28°C for 2–3 h to allow the cells to attach to the dish.
4. Proceed with the infection process using the protocols described in Chapter 6, and after considering sections 9.3 and 9.4 presented below.

9.2.2 LARGE-SCALE CULTURE OF INSECT CELLS IN SUSPENSION CULTURES

For large numbers of insect cells, the use of suspension cell culture is almost mandatory. In the previous section we described how cells which were grown in suspension could be used to inoculate monolayer cultures for subsequent virus infection. The logical progression is to leave the cells as a suspension culture and infect them in this state. There are, however, several problems which must be addressed when considering the large-scale production of insect cells in suspension culture.

Insect cells (Sf) may be grown to a high density in simple spinner cultures of 50–500 ml volumes using flat-bottomed round flasks with a magnetic stirring bar (see Chapter 5). We routinely attain cell densities in uninfected cultures of between $2–3 \times 10^6$ cells per ml. Maiorella *et al.* (1988) reported a density of 5×10^6 cells per ml when using a 100 ml spinner culture. There are problems, however, when one attempts to increase the volume of the cultures beyond 500 ml. In a 3-litre suspension culture, Maiorella *et al.* (1988) could only attain maximum cell densities of 1.2×10^6 per ml. The limiting factor appeared to be the supply of oxygen to the cells, as the ratio of the surface area to the volume of the culture decreases as the size of the culture increases. Oxygen depletion was measured in a closed, liquid filled chamber and calculated at 4.3×10^{-17} moles/cell/s for Sf cells (Maiorella *et al.*, 1988) This is comparable to a 30% reduction in the concentration of oxygen in the air over a 10 min period if there is no replacement. An additional problem is that it has been reported that baculovirus-infected cells have a greater oxygen consumption than non-infected cells (Streett and Hink, 1978). Clearly, if we wish to increase the volume of a suspension culture beyond 500 ml it is necessary to supplement the culture with an additional supply of oxygen or air.

Unfortunately, insect cells are sensitive to shear forces and are damaged by gas sparging (Tramper *et al.*, 1986). This means that if air is sparged through a normal suspension culture many of the cells will be lysed when they come into contact with the small air bubbles. A further complication is that the rising air causes foaming in the culture. Tramper *et al.* (1986) demonstrated that insect cells growing in a fermenter, with a marine-type propeller, were damaged by stirring speeds above 220 revs/min. In bioreactors, efficient dispersal of oxygen or air requires stirring speeds of up to 750–1200 revs/min, so gas sparging at the lower speeds required for insect cells is probably very inefficient. The addition of pluronic F-68 appears to alleviate some of the problems caused by sparging (Maiorella *et al.*, 1988). Despite these problems some success has been achieved with insect cells in large-scale culture in both bioreactors and air-lift fermenters (Miltenberger and David, 1980; Maiorella *et al.*, 1988; Tramper *et al.*, 1990; Weiss *et al.*, 1990).

Establishing a virus infection in suspension cultures of insect cells is more difficult than when using monolayer cultures. There are two options: the first is simply to add the virus to the suspension culture; the second is to concentrate the cells, allow the virus to adsorb and then dilute them to continue the virus replication cycle. The advantage of the first method is that only a single manipulation is required to infect the cells; the disadvantage is that not all of the cells may become infected simultaneously because the virus is necessarily diluted when added to the culture. In the second method, where cells are concentrated before infection, considerably more manipulations are involved, but infection of all of the cells is assured. It is unrealistic, however, to attempt to concentrate very large volumes of cells (greater than about 1000 ml) without specialized cell harvesting equipment that can be maintained in a sterile condition.

A. Infecting suspension cells without concentration

Required:

Suspension of Sf cells which have just attained 10^6 cells/ml
Virus inoculum of known titre

Method:

1. Inoculate the cells with the requisite virus stock at 10 pfu/cell, by simply adding inoculum to the cells.
2. Continue the incubation at 28°C to the optimal time (see Chapter 7) for harvesting recombinant protein.

B. Infecting suspension cells after concentration

Required:

Suspension cell culture as above

Virus inoculum of known titre
TC100 growth medium

Method:

1. Harvest the cells into sterile centrifuge bottles.
2. Pellet the cells at very low speed (about 200 r.p.m.) for 15 min.
3. Resuspend the cells in TC100 medium at a concentration of 10^7 per ml using the appropriate virus inoculum as a component of the diluent.
4. Stir gently at room temperature for 1 h.
5. Dilute the cells to 10^6 per ml in fresh growth medium and continue the incubation at 28°C to the optimal time for harvesting the recombinant protein.

9.3 THE IMPORTANCE OF HIGHLY INFECTIOUS VIRUS STOCKS

In Chapter 6 we described how to prepare large volume, high-titre stocks of recombinant viruses. It is our experience that many workers have difficulty in producing these reagents. While this is not particularly important when performing the initial chracterization experiments illustrated in Chapter 8, you will find the low-titre stocks of virus ($<5 \times 10^7$ pfu/ml) are a distinct handicap when attempting to scale-up protein production beyond 500 ml suspension cultures. This can best be illustrated with the following example, where the aim of the experiment is to infect a 500 ml suspension culture (10^6 cells/ml) with a recombinant virus at a multiplicity of infection of 10 pfu/cell.

Volume of culture:	500 ml
Cell count:	10^6 per ml
Total cell number:	5×10^8
Virus stock:	5×10^7 pfu/ml
Multiplicity of infection:	10
Volume of virus inoculum required:	$(5 \times 10^8 \times 10)/(5 \times 10^7) = 100$ ml

It is obvious that adding 20% of the culture volume as a virus inoculum is unsatisfactory. Furthermore, it is extremely costly of virus stocks and will ensure that almost continuous production of virus inoculum is required. If we look at the same example again, but with a virus stock of 2×10^8, a wholly different result is obtained.

Volume of culture:	500 ml
Cell count:	10^6 per ml
Total cell number:	5×10^8

Virus stock:	2×10^8 pfu/ml
Multiplicity of infection:	10
Volume of virus inoculum:	$(5 \times 10^8) \times 10 / 2 \times 10^8 = 25$ mls

If the virus stock has only a four-fold higher infectivity then the volume of virus inoculum required is reduced to 25 ml; this permits four times the number of protein production runs to be performed with the same volume of virus inoculum. The production of virus stocks with titres greater than 2×10^8 would obviously be even better, but it is comparatively rare to achieve such values.

9.4 MULTIPLICITY OF INFECTION

In the examples discussed in section 9.3 the ratio of infectious virus particles to viable cells (moi) was set at 10. Most virologists will use a moi of between 5 and 50 to infect cells in culture when studying virus replicative processes (e.g. temporal aspects of gene expression, infectious virus production, etc.). This is to ensure that every cell is infected simultaneously and that virus replicative events are synchronous. The higher value is obviously extremely costly in virus inoculum and is only feasible where virus titres of greater than 10^9 are attainable. The problem with using a low moi is that not all of the cells will become infected simultaneously. This presents obvious problems when studying the processes described above. It is also a problem when trying to optimize recombinant protein production by baculovirus vectors in insect cells.

The very late promoters (polyhedrin and p10) used in most of the expression studies to date are active from about 18 to 72 h p.i. Most recombinant proteins, particularly membrane-bound glycoproteins, are not stable indefinitely in virus-infected cells. In consequence, there will be an optimum time to harvest the cells when the concentration of recombinant protein is at its highest (see section 9.5). Unfortunately, this peak will be difficult to ascertain if some cells are infected later than others and, as a result, reach maximum expression at a later time. Therefore, it is of some importance that sufficient virus is used to ensure synchronous infection of all of the insect cells in a culture. We recommend an moi of at least 10 pfu/cell, although as mentioned before it is as well to optimize this figure with your particular virus stock. It must be noted, however, that if you vary the moi, then the time at which the cell is producing maximum yields of virus may also change. An experiment to test the effect of moi on recombinant protein yield must also include an element to monitor the temporal aspects of protein expression. There is no substitute for carefully establishing the conditions most suitable for your own circumstances; we strongly advise against adopting the procedures of other laboratories without confirming that they are also suitable for your own.

9.5 THE OPTIMUM TIME TO HARVEST VIRUS-INFECTED CELLS

The very late baculovirus gene promoters (polyhedrin and p10) used in most of the expression studies are active between about 18 and 72 h p.i. Analysis of the protein concentration in AcMNPV-infected cells shows a very rapid accumulation of material between 24 and 30 h p.i., followed by a more gradual increase over the final part of the infection cycle. Both polyhedrin and p10 proteins are very stable and there is little evidence to suggest that they are subjected to degradation; the polyhedrin protein crystallizes within the nuclei of infected cells to produce polyhedra.

In cells infected with recombinant viruses, containing foreign coding sequences under the control of the polyhedrin and p10 promoters, it is unwise to assume that the foreign gene product will

Table 9.1 Examples of proteins that have been purified from insect cells infected with recombinant baculoviruses

Protein	*Major purification step(s)*	*Reference*
BTV-10 NS1 gene product	Gradient centrifugation	Urakawa and Roy (1988)
BTV-10 VP7 gene product	ammonium sulphate fractionation and ion-exchange chromatography	Oldfield *et al.* (1990)
HBV core antigen	CsCl density gradient centrifugation	Hilditch *et al.* (1990)
HBV surface antigen	Detergent extraction and immunoaffinity chromatography	Landford *et al.* (1989)
HSV-1 helicase-primase	DEAE-sepharose chromatography	Dodson *et al.* (1989)
HSV-1 DNA polymerase	Phosphocellulose chromatography	Marcy *et al.* (1990)
Human CD4 receptor	Detergent extraction and immunoaffinity chromatography	Webb *et al.* (1989)
Protein phosphatase 1	DEAE-sepharose chromatography	Berndt and Cohen (1990)
Rabies virus nucleoprotein	Differential centrifugation and ion-exchange chromatography	Préhaud *et al.* (1990)

be as stable as the native virus proteins. It is more likely that once the period of peak synthesis has been passed, the steady-state levels of recombinant protein will decline as continued production is unable to match the degradative processes. The optimum time for harvesting the infected cell culture should be determined experimentally.

9.6 PURIFICATION OF RECOMBINANT PROTEIN FROM INFECTED CELL CULTURES

It is beyond the scope of this book to consider the many options available for extracting the recombinant product from insect cells. The reader should consult other texts for detailed protocols. In essence, insect cells may be considered in the same category as vertebrate cells when it is necessary to purify proteins. They may be easily lysed in non-ionic and ionic detergents or readily disrupted using sonication apparatus.

Table 9.1 lists some examples of recombinant proteins that have been purified from infected insect cells. These may provide some ideas as to the most suitable procedures for your own experiments.

Propagation of baculoviruses in insect larvae 10

10.1 INTRODUCTION

This book is concerned with the use of baculovirus expression vectors to produce foreign proteins in insect cell culture. Historically, interest in baculoviruses was focused on their growth in insect larvae because of the potential of the viruses as biological insecticides to control pests in agriculture and sericulture. As described in Chapter 1, the advances in genetic manipulation of baculoviruses now permit the modification of the virus insecticides to improve their efficacy. DNA sequences encoding insecticidal toxins, hormones or enzymes may be inserted in the baculovirus genome so that on expression in the insect larva, the recombinant protein hastens its demise. In this chapter we describe some basic protocols required for the propagation of baculoviruses in insect larvae. These are not intended to be a comprehensive guide to the subject area, but to serve as an introduction to the non-specialist. We hope to stimulate interest in an area of baculovirus research that offers considerable potential for improving virus insecticides. Although most projects with this goal adopt a rational approach to the selection of foreign proteins for production in insect larvae, it is likely that some of the recombinant products currently being made by other groups, for other purposes, could have an insecticidal effect, particulary secretory proteins. This chapter should demonstrate that the testing of recombinant viruses in insects is not such a major undertaking as it might at first appear.

The other role for this chapter is to suggest the use of insects, in lieu of cell culture, for the production of recombinant protein. This is an option which has received scant attention in previous reviews. Considering the yield of wild type virus in infected larvae (approximately 10^9 polyhedra/larva), it is surprising that little has been done to utilize this extremely efficient *in vivo* bioreactor to produce foreign proteins. The most suitable virus–host combination to use for this purpose is the BmNPV and the *B. mori* (silkworm) larva. This expression system has been described by Maeda *et al.* (1985) and has been recently reviewed by the same author (Maeda, 1989a). The long history of culturing this insect for silk production has selected large larvae which are easily handled using semi-automated methods. Unfortunately, most users of the baculovirus expression system are more familiar with the AcMNPV system. In consequence, this virus must be used with other insect hosts such as *Trichoplusia ni*, which is also easily reared in laboratory conditions. A complication for most users of the baculovirus system is that the recombinant viruses produced for expressing the foreign proteins will be of the polyhedrin-negative phenotype. This is an inevitable result of replacing the polyhedrin coding sequences with those of the foreign gene. For reasons which are not fully understood, such viruses are relatively inefficient at establishing an infection in insect larvae through the natural infection route (*per os*). Instead, virus particles must be injected directly into the haemolymph, a time-consuming and delicate operation. This problem may be circumvented by constructing recombinant viruses which retain a fully functional polyhedrin gene, in addition to the foreign coding sequences. This is achieved by making use of the p10 gene locus for inserting the foreign sequences (Vlak *et al.*, 1988; Williams *et al.*, 1989; Weyer *et al.*, 1990) or by duplicating the p10 promoter upstream of the polyhedrin gene (Weyer *et al.*, 1990). The vectors appropriate for this approach (pAcUW1 and pAcUW2B) were described in Chapter 2. Polyhedrin-positive recombinant viruses may be extracted, as polyhedra, from infected cells *in vitro* and then used efficiently to infect insect larvae by direct feeding on inoculated diet.

10.2 REARING INSECTS IN THE LABORATORY

It is recommended that advice is sought from experts in entomology if you wish to set up a colony of insects in the laboratory (Insectory). It is impossible to provide sufficient details here to enable successful initiation of such a facility. We can only give some initial guidelines and suggestions as to the equipment and labour required. For some insects which are not native species it is necessary to obtain a licence from the relevant authorities (The Ministry of Agriculture, Fisheries and Food, in the UK).

With many species it is now possible to rear the larvae on a semi-synthetic diet (see section 10.8), which obviates the need for fresh foliage (cabbages, etc.). This also means that it is quite simple to maintain the laboratory colony throughout the year, the only proviso being that the diapause of the insect can be easily broken. With relatively simple facilities, a constant temperature room or incubator, a cold room or refridgerator, it is possible to produce many thousands of insects on a weekly basis at comparatively low cost. The labour required is minimal; the insectory at IVEM, Oxford has one full time member of staff, assisted by a part-time worker to tend about fifteen insect species. This facility is housed in an area which is kept virus-free. Additional precautions, such as surface sterilization of eggs between generations, ensures that the population does not suffer an unplanned virus outbreak. Matings of adults are timed to ensure that larvae are in constant supply.

For the protocols described in this chapter we will be referring to AcMNPV and *T. ni* larvae as examples. Most other virus–host combinations will work in the same way.

10.3 INFECTION OF INSECT LARVAE WITH POLYHEDRA FROM CELL CULTURE

Establishing an infection in insect larvae is a simple matter of feeding polyhedra to the larvae in the form of contaminated diet and then allowing the insects to consume virus-free diet until death. The only complication is that it is necessary to use insects at an early stage in their larval development because the virus dose required to initiate infection increases dramatically as the host increases in size. The LD_{50} for AcMNPV in early third instar *T. ni* is about 100 polyhedra; this increases to several thousand polyhedra in the fourth and fifth instars. Since the starting material for infecting the insects will be polyhedra from infected cells in culture, it is essential to conserve this material to avoid having to prepare large quantities in suspension culture. Early larval instars are therefore used to propagate the virus.

10.3.1 PREPARATION OF VIRUS POLYHEDRA FROM INFECTED CELLS IN CULTURE

It is not necessary to purify polyhedra from virus-infected cell cultures subsequently to infect insect larvae. Small quantities of Sf cells may be infected, allowed to synthesize polyhedra, harvested and then fed directly to the larvae, by placing the intact cells on plugs of diet, to initiate infection. It seems that the extremely alkaline conditions of the larval mid-gut are sufficient to lyse the infected cells and release polyhedra into the lumen of the gut.

Required:

1×25 cm^2 flask seeded with 1×10^6 Sf cells

TC100 growth medium
Recombinant virus (polyhedrin-positive)
PBS

Method:

1. Check that the cells are sub-confluent before use. Remove the medium and inoculate the cells with about 100 μl virus (0.1 pfu/cell). Incubate at room temperature for 1 h. Remove the virus inoculum and add 4 ml growth medium. Incubate at 28°C until all cells have polyhedra within the nuclei.
2. Harvest cells and medium together. Pellet the cells using low-speed centrifugation and discard the supernatant fraction to disinfectant (alternatively, retain the medium as a stock of infectious virus particles for future use).
3. Resuspend the cell pellet in a small volume of PBS. The final volume should be 100–200 μl. Store at 4°C in the short term and at −20°C for longer periods (months/years).

10.3.2 PROPAGATING THE VIRUS IN INSECT LARVAE

Required:

Early third instar *T. ni* larvae
Recombinant baculovirus as polyhedra (see section 10.3.1)
96-well microtitre plates
Semi-synthetic diet (see section 10.8)
Paint brush or blunt forceps, for handling larvae
Polypots (Ashwood Timber and Plastics Ltd) containing diet
PBS

Method:

1. Place small (ca. 1 mm^3) pieces of diet in each well of a 96-well microtitre plate. Inoculate each portion of diet with 1 μl of the infected cell extract (undiluted, 10^{-1} and 10^{-2} dilutions, in PBS).
   * The dose of virus given to the insects at this stage is not critical.
   * If a defined stock of virus polyhedra is available, dilute the suspension in water to a concentration of 500 polyhedra/μl, then inoculate each diet plug. This procedure is generally used when passaging the virus for a second time in the insect larvae. The method for purifying polyhedra from insects is described in section 10.4.
2. Place an insect larva in each well of the microtitre plate and cover with a small piece of glass to prevent the insects escaping.
   * Handle the insects very carefully with blunt forceps or a paint brush.

3. Place the lid in position and incubate in a moist atmosphere for 24 h at 21–28°C.
4. Observe the insects and note those which have consumed the contaminated diet. Transfer these individuals to small plastic containers (e.g. 'polypots'), with sufficient diet (see section 10.8) to support the insects for the remainder of their feeding period (about 0.4 cm depth in the polypot).
5. Incubate as described in point 3 until the larvae succumb to virus infection or pupate. Harvest the insects when they become creamy white (an example of an infected larvae is shown in Figure 1.3); avoid the onset of bacterial growth which turns the larvae black and results in the contamination of the polyhedra preparation. Store the virus-infected cadavers at −20°C until required.

   * The above procedure is very economical of virus inoculum and semi-synthetic diet. It should be used in the first round of virus replication in insects. Larger batches of virus may be grown by contaminating the surfaces of diet, dispensed in large Petri dishes (or 'margarine tubs'), with the polyhedra and then allowing several larvae to feed on each dish. The virus dilution may be spread in a manner similar to the inoculation of agar plates with bacteria. This method requires more virus inoculum, but avoids the need to handle the insects more than once in the infection process. Virus-infected insects are picked off the diet when appropriate.

10.4 PURIFICATION OF POLYHEDRA FROM INFECTED LARVAE

The purification of polyhedra is complicated by the presence of considerable quantities of larval debris in the starting material. This must be removed prior to the fractionation of the polyhedra using sucrose gradients. Despite the rather unpleasant appearance of the infected insects, it is possible to derive very clean preparations of polyhedra which are very largely free of other materials.

Required:

Muslin
Sucrose gradient maker
30% (w/w) sucrose in TE (note w/w and not w/v)
50% (w/w) sucrose in TE (note w/w and not w/v)
60% (w/w) sucrose in TE (note w/w and not w/v)
0.1% w/v SDS
TE (10 mM Tris-HCl, pH 8.0, 0.1 mM EDTA)

Macerator e.g. Colworth Stomacher (useful for very large batches of infected insects)

Precautions:
Only the very strongest disinfectants will inactivate polyhedra (e.g. Chloros, hypochorite or a strong bleach). The use of the autoclave is often more convenient. While the polyhedra do not present a risk to humans, contaminated work surfaces or equipment may lead to unwanted infections in insect larvae or in cell culture. If at all possible, work in a Class I safety cabinet or fume hood to avoid spread of the virus-contaminated materials used in the purification to other parts of your laboratory.

Method:
1. Thaw virus-infected insect larvae from 10.3 at 37°C.
2. Place the insects in a stout plastic bag and add a little 0.1 % SDS solution.
3. Thoroughly macerate the cadavers until a smooth paste is produced. It may be necessary to add extra 0.1% SDS if the mixture is too solid.
4. Place two layers of muslin in a glass funnel and mount in a clamp and stand with a conical glass vessel positioned beneath the funnel.
5. Filter the larval macerate through the muslin. Add extra 0.1% SDS to the mixture in the funnel to ensure that all of the virus polyhedra are removed.
6. Discard the muslin and residual portions of the insects by autoclaving.
7. Pellet the polyhedra from the filtrate at 3000 *g* for 30 min.
8. Add approximately 10 volumes of TE to the pellet and resuspend the polyhedra. Transfer to a glass universal bottle and stir at room temperature for 30 min to ensure thorough dispersal of polyhedra aggregates. Add more TE if the suspension appears too concentrated; it is better to overdilute the preparation at this stage.
9. Prepare 50–60% (w/w) discontinuous sucrose gradients. These gradients consist simply of a layer of 50% (w/w) sucrose over a 60% (w/w) sucrose cushion. The size required will depend on the quantities of virus being processed.
   * Since the volume of crude polyhedra will be quite large it is often necessary to leave about half of the tube available for loading the virus. This means that the gradient will be quite small; this is not a problem since this first fractionation is only intended to remove the greater proportion of the unwanted insect debris.

10. Carefully load the crude polyhedra suspension onto the sucrose gradient; overlay with liquid paraffin to balance all of the tubes.
11. Spin the gradients at 90 000 *g* for 1 h to band the polyhedra at the interface between the two sucrose layers. Heavy debris from the insect forms a considerable pellet at the bottom of the tube.
12. Harvest the virus by puncturing the side of the centrifuge tube (just above the pellet of debris) and collecting the appropriate portion of the sucrose gradient. This operation may be regulated by inserting a rubber bung, pierced through with a 19G needle attached to a flexible tube, in the mouth of the centrifuge tube.
    * Harvesting the virus polyhedra by using a large needle and syringe inserted through the side of the tube is feasible but often complicated by the extremely viscous nature of the polyhedra suspension.
    * It is also possible to harvest the band of polyhedra from above using a Pasteur pipette.
13. Dilute the virus suspension at least 1:3 using TE and load onto 30–60% (w/w) continuous sucrose gradients. Overlay with liquid paraffin as before.
    * Use a gradient maker to prepare the continuous sucrose gradients.
14. Spin at 90 000 *g* for 1 h. In this continuous gradient the polyhedra form a more diffuse band.
    * This second gradient serves to separate the insect debris with a similar sedimentation rate to the polyhedra.
15. Harvest the polyhedra as described in 12.
16. Dilute the polyhedra approximately 1:10 and pellet at 10 000 *g* for 30 min.
17. Discard the supernatant fraction and add approximately 10 volumes of TE to the pellet of polyhedra and soak overnight at 4°C. This serves to soften the virus pellet and aid subsequent resuspension. The final concentration of this preparation should be about 5–10 mg/ml. The polyhedra may also be counted using a haemocytometer.
18. Store the polyhedra at 4°C or −20°C.
    * Polyhedra stored at −20°C tend to aggregate after thawing, therefore storage at 4°C is recommended. If, however, there is bacterial contamination of the preparation, then storage at −20°C may be necessary to avoid further growth of the micro-organisms.

10.5 BIOASSAYS OF POLYHEDRA

Two major parameters are usually measured after feeding purified polyhedra to insect larvae. The first concerns the number of polyhedra required to kill the host; usually expressed as the lethal dose $(LD)_{50}$ (i.e. the dose required to kill 50% of the population). The second determines the time required to kill the host; this is usually expressed as the lethal time $(LT)_{50}$.

The methods described here are quite simple to perform, but we recommend once again that assistance is obtained from experienced scientists. It is impossible to convey all of the subtle points required to conduct statistically reliable bioassays in the space available in this chapter.

10.5.1 LD^{50} ASSAYS

Required:

Second instar *T. ni* larvae (preferably from a single hatch in a 12 h period)
Recombinant baculovirus as purified polyhedra (section 10.4)
Wild type AcMPNV as purified polyhedra (control)
96-well microtitre plates
Semi-synthetic diet (see section 10.8)
Paint brush or blunt forceps for handling larvae
Polypots (Ashwood Timer and Plastics Ltd) with diet

Method:

1. Prepare dilutions of the virus(es) to be assayed by diluting polyhedra in water. Polyhedra may be counted using a standard counting chamber, e.g. a haemocytometer. The maximum dose of the virus should ideally give over 90% mortalities and the minimum dose should give 10% mortalities. Non-engineered AcMNPV fulfil these conditions with a dose range of 120, 60, 30, 15 and 7.5 polyhedra per larvae. The virus dose should be delivered in 1 μl volumes.
2. For each virus dose, place 50 small (1 mm^3) plugs of semi-synthetic diet into the wells of a 96-well microtitre dish. Inoculate each plug of diet with 1 μl of the required virus dose.
3. Place a larvae into each well containing diet, cover the tray with moist tissue paper and a glass plate and incubate at 22°C for 24 h.
4. Those larvae that have fully consumed the plug of diet should be transferred to fresh diet in a polypot (1 larva per pot). The remaining insects should be discarded.
5. Monitor the insects for signs of infection on a daily basis, until

the larvae pupate. At this point record the numbers that have been infected.

6. Calculate the LD_{50} value using standard statistical programs.

10.5.2 LT_{50} ASSAYS

Required:

Neonate *T. ni* larvae
Recombinant baculovirus as purified polyhedra (section 10.4)
Wild-type AcMNPV as purified polyhedra (control)
Food dye (blue)
Semi-synthetic diet (see section 10.8)
Paint brush or blunt forceps for handling larvae
Polypots (Ashwood Timber and Plastics Ltd) with diet

Method:

1. After hatching, starve neonate larvae for 3–6 h.
2. Feed the larvae with a virus suspension (2×10^6 polyhedra/ml) coloured with a blue food dye (5% solution). This is best achieved by placing small droplets of virus suspension onto a Petri dish in concentric rings. The larvae are introduced into the centre of the rings, after which they move through the droplets, taking in a small volume of liquid, before crawling onto the Petri dish lid. This procedure allows synchronous feeding and the selection of healthy individuals.
3. Transfer neonates from the Petri dish lid to polypots of diet (1 per pot) and monitor daily for signs of infection and death.
4. Calculate the LT_{50} values using standard statistical programs.

10.6 PURIFICATION OF VIRUS PARTICLES AND DNA FROM POLYHEDRA

Very substantial quantities of polyhedra may be purified from infected larvae. These may be used as source of virus DNA for subsequent transfection experiments when producing recombinant viruses. Milligram quantities of DNA may be prepared when using insect-derived polyhedra as the starting material. This amount of virus DNA would satisfy the requirements of most laboratories for several years! The use of insects to produce virus DNA is, therefore, an option well worth serious consideration.

Required:

30% (w/w) sucrose in TE
60% (w/w) sucrose in TE
Gradient maker

TE (10 mM Tris-HCl, pH 8.0, 0.1 mM EDTA)
0.1 M sodium carbonate
Reagents for DNA purification described in Chapter 6

Method:

1. Mix an equal volume 0.1 M sodium carbonate with a suspension of polyhedra and stir at 28°C for 30 min. The suspension should become less opaque as the polyhedra dissolve. A milky white appearance will remain as the concentration of virus particles is very high.
2. Pellet the undissolved polyhedra at 3000 *g*. This serves as a good check on the efficacy of the dissolution procedure. If the pellet is substantial, resuspend in water and repeat the dissolution process described in point 1.
3. Prepare one or more 30–60% (w/w) continuous sucrose gradients and layer the virus particle suspension on top. Overlay with liquid paraffin.
4. Spin at 90 000 *g* for 1 h at 4°C. Several virus bands will be observed in the final gradient, as shown in Figure 10.1. This provides a graphic illustration of the variable numbers of nucleocapsids packaged within each of the virus particles.
5. Harvest the virus bands by piercing the bottom of the tube (see section 10.4) and pool.
6. Dilute the virus suspension 1:3 with TE and pellet the virus particles at 90 000 *g* for 1 h.
7. Discard the supernatant fraction and soak the virus pellet in a small volume of TE at 4°C overnight. Resuspend the virus and store at 4°C.
8. To extract infectious DNA from the purified virus, refer to Chapter 6.

10.7 ISOLATION OF VIRUS PARTICLES FROM INFECTED LARVAE TO ESTABLISH INFECTIONS IN CELL CULTURE

While most work on baculovirus molecular genetics and analysis of virus gene expression is now performed in cell cultures, the original virus inoculum was derived from insect larvae. There are two options for deriving infectious material subsequently to establish an infection in cell culture. In the first, polyhedra are dissolved in an alkaline buffer to release infectious virus particles. In the second, polyhedra are fed to insect larvae which are then bled to obtain haemolymph containing infectious virus (ECV). While the first method may appear more attractive, it suffers the disadvantage that the virions produced are at least 1000-fold less infectious for cells

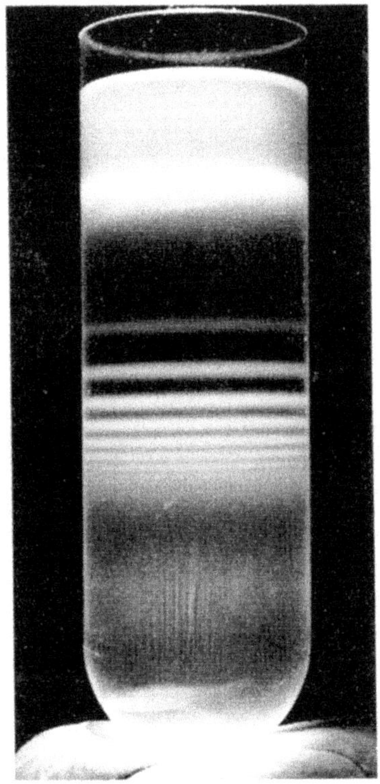

Figure 10.1 Purification of virus particles from insect-derived polyhedra. Polyhedra were incubated for 30 min at 28°C in the presence of 50 mM calcium carbonate to dissolve the polyhedral protein and release virus particles. The liberated virus particles were subsequently fractionated in a 30–60% (w/w) sucrose gradient at 90 000 *g* for 1 h. The presence of multiple bands in the gradient indicates the presence of variable numbers of nucleocapsids in the virus particles. We thank Martin Ayres for supplying this figure.

in culture than the virus particles derived from haemolymph. This is because the virus in haemolymph is produced by budding from the infected cells lining the haemocoel; in consequence the particles have gp67 in the virus envelope. The virus particles released from polyhedra lack gp67 because they were originally occluded within the nuclei of infected cells. For a fuller description of this area see Chapter 1. For AcMNPV both methods are successful in deriving infectious material for initiating virus replication in cell culture. In other virus systems, only the haemolymph-derived material is able to perform this role, presumably because the cell lines used are only semi-permissive for the virus (e.g. *M. brassicae* MNPV and *M. brassicae* cells).

10.7.1 PURIFICATION OF VIRUS PARTICLES FROM POLYHEDRA FOR THE INFECTION OF CELLS IN CULTURE

Required:
- 0.1 M sodium carbonate
- 100 mM Tris-HCl, pH 7.5
- 0.45 μM disposable filters
- Polyhedra (see section 10.4)
- Sf cells (in 25 cm^2 flasks; see Chapter 5)

Method:
1. Mix an equal volume of 0.1 M sodium carbonate with a suspension of polyhedra and stir at 28°C for 30 min.
2. Adjust the pH of the suspension to less than 8.0 by adding 100mM Tris-HCl, pH 7.5 and pass through a 0.45 μM disposable filter.
3. Use dilutions of the filtrate to inoculate Sf cells as described in Chapter 6.
4. Monitor the infection by the appearance of polyhedra. Harvest the cell culture medium to provide a stock of virus for subsequent propagation *in vitro* (refer to Chapter 6).

10.7.2 PURIFICATION OF VIRUS PARTICLES FROM HAEMOLYMPH FOR THE INFECTION OF CELLS IN CULTURE

Required:
- 3rd instar *T. ni* larvae
- Semi-synthetic diet (see section 10.8)
- Polypots (see section 10.5)
- 0.45 μM disposable filter
- Polyhedra (see section 10.4)
- 96-well microtitre plates
- Sf cells (in 25 cm^2 flasks, see Chapter 5) and TC100/10% FCS
- 0.1 M cysteine (optional; prepare fresh each time)

Method:
1. Infect individual *T. ni* larvae (50–100) with 1000 polyhedra each as described in section 10.3.2 using the microtitre plate procedure.
2. Transfer the larvae to individual polypots and incubate at 28°C for 4–5 days. If the larvae show signs of infection at this stage then the process may have proceeded too far to derive adequate quantities of haemolymph. In this case, repeat the infection with fresh larvae using a lower dose of virus.

3. Surface sterilize the larvae by immersing in absolute alcohol for 1 min.
   * This procedure also serves to immobilize the larvae and considerably eases the harvesting of haemolymph.
4. Place each larva on a sterile Petri dish and nick a leg with a sterile surgical blade. Application of gentle pressure on the insect will increase the flow of haemolymph from the wound. Collect the haemolymph with a needle and syringe containing about 0.5 ml TC100/10% FCS.
   * Take care not to pierce the gut when performing this operation.
   * At this stage if melanization (blackening) is a problem, the addition of cysteine to 1 mM is very helpful in slowing its progress.
5. Repeat step 4 for each larva (50–100 in total); use the same syringe for harvesting the haemolymph to provide a substantial pool of infected material.
6. Filter the haemolymph using a 0.45 μM disposable unit, to remove bacterial contamination.
   * Although it is possible to store this infectious material at 4°C for a few days, or at −70°C for several weeks, it is advisable to use it as soon as possible. The insect haemolymph is rich in proteases and these will eventually inactivate the virus.
7. Inoculate Sf cells in 25cm^2 flasks with dilutions of the virus-contaminated haemolymph (undiluted, 10^{-1} to 10^{-3}). Incubate at room temperature for 1 h, discard the virus inoculum and replace with 5 ml TC100 growth medium.
8. Incubate at 28°C until the cells contain polyhedra. Harvest the medium, remove the cells using low-speed centrifugation, and store the supernatant fraction at 4°C or −70°C until required.
   * The medium becomes a source of virus inoculum for further infections of insect cells in culture.

10.8 PREPARATION OF SEMI-SYNTHETIC INSECT DIET

Required:

8 oz plastic tubs with clear lids (e.g. 'margarine tubs')
and/or polypots (Ashwood Timber and Plastics Ltd)
Kenwood Chef or other electric food mixer with detachable whisks that are completely autoclavable (i.e. no plastic parts)
or
Wooden spoon, hand-held metal whisk and large, autoclavable polypropylene beaker
Ingredients as listed in points 2, 6 and 9 below

Linseed oil
Access to a laminar flow hood

Method:

1. Sterilize the whisk attachments from the electric mixer by autoclaving. Alternatively, if you are only making relatively small amounts of diet, sterilize a wooden spoon and metal hand-held whisk.
2. Place the following ingredients into the mixer bowl or a large autoclavable beaker (not glass). You can pour approximately 45 8 oz pots and 60 polypots from 1 l diet.

   Diet recipe:

	g/l diet
Bactoagar (Difco)	20
Casein (Sigma)	35.2
Wheatgerm (e.g. Beemax)	76.8
Wessons salts (see point 9 below)	10
Dried bakers' yeast	15.2
Sucrose	31.2
Cholesterol (Sigma)	1
Methyl-4-hydroxy-benzoate (Sigma)	1
Sorbic acid (Sigma)	20

3. Mix to a paste with 1013 ml water and add 2 ml linseed oil.
   * Linseed oil can be obtained from chemist or sports shops; used for oiling cricket bats!
4. After mixing well, cover with foil and autoclave for 20–30 min at 15 p.s.i. (the volume will be slightly reduced)
5. On removing from the autoclave, place in a laminar flow hood and stir well with a wooden spoon/metal whisk or use an electric whisk, until the temperature falls to 60–70°C.
6. Mix in the following, stirring for about 5 min (by hand) or 2 min using the electric mixer:

	g/l diet
Vitamin/antibiotic mixture (see point 9 below)	6
Choline chloride (Sigma)	1

7. Quickly, before the agar starts to set, pour the diet into polypots or 8 oz pots, as required. Allow to set, place on lids and store at 4°C. Use within 1–2 weeks.
8. When using diet pots for insect rearing, remember to cut the polypot or 8 oz tub lids with a scalpel blade, to allow exchange of gases and to permit water vapour to escape.

9. Lists of ingredients for Wessons salts and vitamin/antibiotic mixture

Wessons salts:
Mix all ingredients together very well (this is very important and you may need to use a mortar and pestle) and store in a dry, dark bottle at room temperature. All salts should be of Analar quality.

$CaCO_3$	60g
$K_2HPO_4.3H_2O$	84.5g
$CaHPO_42H_2O$	15g
$MgSO_4.7H_2O$	20.4g
NaCl	33.5g
$FeC_6H_5O_7.6H_2O$	5.5g
KI	0.16g
$MnSO_4.4H_2O$	1.0g
$ZnCl_2$	0.05g
$CuSO_4.5H_2O$	0.06g

Vitamin stock:
Mix together (very well) the following ingredients and store at 4°C in a dry, air-tight bottle. All vitamins can be purchased from Sigma.

Nicotinic acid	5.0g
Ca-pantothenate	5.0g
Riboflavin (vitamin B2)	2.5g
Thiamine (vitamin B1)	1.25g
Pyridoxine HCl (vitamin B6)	1.25g
Folic acid	1.25g
D-Biotin	0.1g
Cyanocobalamin (vitamin B12)	0.01g

Antibiotic/vitamin mixture:
Prepare this mixture on the day required or the day before and store at 4°C overnight. Do not add to diet until it has cooled to 60–70°C.

Vitamin mixture (see above)	0.26g
Streptomycin (Sigma)	0.53g
Ascorbic acid (vitamin C) (Sigma)	10.66g
Aureomycin (veterinary powder)*	4.8g

*Aureomycin can be purchased from your local veterinary surgeon.

Trouble-shooting guide 11

11.1 INTRODUCTION

In the various sections throughout this book we have endeavoured to highlight areas which may cause difficulties to the first-time or even experienced user of the baculovirus expression system. Here, we have summarized these points to provide a simple problem-solving guide.

11.2 INSERTION OF FOREIGN GENE CODING SEQUENCES INTO TRANSFER VECTORS

Difficulties at this stage are not peculiar to the baculovirus system. Any technique which involves inserting foreign DNA into a plasmid vector will encounter problems on occasion. These may be a consequence of faulty reagents (plasmids, enzymes, etc.) or the particular secondary structure of a DNA fragment which renders it difficult to insert into the chosen restriction enzyme (RE) site.

11.2.1 THE TRANSFER VECTOR

A stock of the baculovirus transfer vector should be prepared which is of the highest quality. This will usually involve one or even two equilibrium gradients with caesium choride. Alternative methods of preparation are now available which may produce comparable results, such as chromatographic techniques and mini-columns. Whichever method is used the DNA must appear clean on agarose gel (no chromosomal DNA) and have a good 260/280 nm ratio of about 1.6–1.8. It must also be readily cleavable with REs, as judged by agarose gel electrophoresis and reduction in transformation efficiency. If the transfer vector is incompletely digested then a very high background of bacterial colonies after transformation of *E.*

coli will mask the true recombinants. Incomplete dephosphorylation of the digested plasmid will also result in the same problem. If it is envisaged that a large number of expression experiments are to be undertaken then it is well worth preparing a large stock of RE-digested, dephosphorylated transfer vector and storing it at −20°C until required. The time and care invested at this early stage of the expression project will be rewarded with fewer problems when trying to insert the foreign DNA.

If the foreign gene is relatively large it may be difficult to insert it into the larger transfer vectors (e.g. pAcRP23, pAcYM1, see Chapter 2). When we have had this problem it has usually been overcome by using one of the newer, smaller transfer vectors (e.g. pAcAL1, pAcAL2, see Chapter 2).

11.2.2 DNA SEQUENCES FOR INSERTION IN TRANSFER VECTORS

The same general guidelines described above (section 11.2.1) apply to the preparation of plasmids containing the foreign coding sequences for inserting into the baculovirus transfer vectors. It is possible to use mini-prep DNA as a source for isolating the fragment of interest, but this is more likely to cause to problems than if a carefully purified maxi-prep is used.

When running the preparative gel after digestion of the plasmid, always estimate in advance where the desired DNA fragment will migrate relative to the plasmid backbone. If these components are uncomfortably close together in the gel, then digest the plasmid sequences with another RE to increase the separation. Excessive contamination of the foreign gene coding sequences with the plasmid backbone will give a high background in the subsequent ligation/transformation stages. We have noted that a higher proportion of the plasmid backbone co-purifies with the required DNA fragment if the latter is the larger of the two. Always estimate the concentration of the DNA fragment after purification with the aid of an agarose mini-gel and appropriate DNA size markers.

A common concern is whether the nature of the 5′ and 3′ untranslated sequences attached to the foreign gene coding region will adversely affect subsequent expression in the baculovirus–insect cell system. To date, there is little evidence to suggest that these sequences have such an effect. The length of each of these regions does not appear to correlate with the level of protein expression. Unfortunately, a good controlled study has yet to be done where the nucleotide composition and length of the non-coding regions are varied prior to the testing of expression in insect cells. However, it is very important that there are no ATG initiation sequences upstream of the authentic one. If such a sequence is present, it must be removed, e.g. by site-directed mutagenesis.

A particular concern is often whether the foreign gene has an appropriate Kozak consensus sequence (Kozak, 1986) for translation initiation. Once again, a detailed study of this factor has yet to be published, but is our experience that provision of the ideal Kozak sequence before the foreign coding region does not guarantee a high level of expression. There is no evidence to suggest that codon usage affects the efficiency of translation, although if you are synthesizing a gene from scratch, it may be worthwhile using the preferred codons for the polyhedrin gene.

In summary, we recommend that it is not worth paying too much attention to the structure of the foreign DNA to be inserted into the baculovirus expression vector. Given the rapid methods now available for the construction of a recombinant virus (see Chapters 2 and 7), it is probably better to test protein synthesis with the foreign coding region in its native form, rather than investing too much effort in tailoring the DNA fragment. If PCR technology is available in your laboratory, however, it may be worthwhile to use this versatile method to tailor the foreign gene before insertion into the transfer vector. Subsequent modification of the foreign DNA sequences is also assisted if the baculovirus transfer vector has the facility to produce single-stranded DNA (e.g. pAcCL29 and pAcAL1, see Chapter 2). This may be used with site-directed mutagenesis methods rapidly to change the region of interest.

11.2.3 LIGATIONS

Although many protocols have been described in the literature, this relatively simple technique can still be a source of considerable frustration and delay in the expression project. If performed carefully then valuable information may be obtained, even when the experiment does not appear to have been successful. With the transfer vectors currently in use there is no easily selectable phenotypic marker, such as β-galactosidase or inactivation of antibiotic resistance, to facilitate detection of plasmids with inserts. Instead, it is necessary to rely on the difference in colony numbers between ligation reactions containing the transfer vector and insert DNA and those reactions lacking the insert. In summary, the following ligation reactions should always be set up.

1. Digested transfer vector
2. Digested transfer vector plus ligase

These serve as a control for the efficiency of the ligation reaction.

3. Digested transfer vector/CIP treated
4. Digested transfer vector/CIP treated, plus ligase
   These serve as a control for the efficiency of dephosphorylation. (Reaction 3 need only be done the first time that the transfer vector is used.)
5. Digested transfer vector/CIP treated, plus ligase, with variable concentrations of insert DNA
   The ratio of colonies produced between reactions four and five will give an estimate of the efficiency of the insertion of the foreign coding sequences into the transfer vector.
6. Purified insert sequences plus ligase
   This control may appear superfluous, but it does provide an estimation of the number of colonies in reaction 5 which are a consequence of contamination with plasmid backbone sequences.

A problem occasionally encountered is that the foreign gene coding sequences can only be inserted into the transfer vector in the wrong orientation. Screening numerous colonies using filter-hybridization techniques will eventually identify the appropriate plasmid. Alternatively, the baculovirus transfer vector and foreign DNA may be modified to permit 'forced cloning' of the insert via the use of asymmetric RE sites.

11.3 CELL CULTURE

Some problems that may be encountered and possible solutions are listed below. A good indication that cells are not thriving well or that have been left too long between passages is the appearance of 'sausage' shaped cells that have arrested during cell division.

11.3.1 CELLS FAIL TO THRIVE AND ATTACH TO GLASS/PLASTIC SURFACES

(a) Has the foetal calf serum been added? If making media from basic ingredients, have any been accidently omitted?
(b) Inspect cells under high power on the microscope. Can you see polyhedra, indicating contamination with AcMNPV, or cytopathic effect due to polyhedrin-negative recombinant viruses? If so, dispose of cells by autoclaving and retrieve fresh cells from other stock flasks or from liquid nitrogen. Review cell culture procedures to eliminate contamination.

(c) If using glass flasks, are the washing and rinsing procedures removing all of the detergent? See Chapter 5 for recommended washing procedures.
(d) Is there a mycoplasma contamination? Mycoplasma contamination is easy to miss in the early stages, but if cells fail to grow, fail to support virus replication and have a grainy appearance with motile organisms moving within the cell, then this may be the cause. Kits to detect mycoplasma are commercially available (e.g. from ICN-Flow), which are based on the Hoerst stain. A number of antimycoplasma agents are now on the market, but none that we have tried appear to work with insect cells. It is good practice to screen all cells received from other laboratories for possible contamination. Prevention seems better than cure in this case.
(e) Is the passage number too high? If so, retrieve fresh stocks from liquid nitrogen.

11.3.2 CELLS ARE CONTAMINATED WITH VIRUS

See section 11.3.1 (b).

11.3.3 CELLS ARE CONTAMINATED WITH YEAST, FUNGI OR BACTERIA

(a) Has the medium been sterility checked after adding FCS? If microbial contamination becomes a problem it may be best routinely to check bottles of media after adding FCS.
(b) Is the water bath that is used to warm media or heat-inactivate FCS clean? Water baths are often a source of *Pseudomonas*, a common contaminant that turns insect cell cultures a creamy or green colour.
(c) Are stock bottles of media, etc. swabbed with alcohol after long-term storage at 4°C?
(d) Are other people culturing bacteria, yeast or fungi nearby? If so, you may need to review containment procedures.

11.3.4 CRYSTALS OF PRECIPITATE IN THE MEDIUM

(a) This may sometimes occurs with 'home-made' TC100. The reasons for the crystals appearing are unclear but they are probably due to the pH of the medium rising above 6.3, and thereby forming an insoluble precipitate. Crystals are very difficult to remove from cells and in our experience it is best to re-pH and filter the medium and retrieve fresh cells from liquid nitrogen or other stocks not containing crystals.

11.4 PREPARATION OF VIRUS STOCKS AND INFECTIOUS DNA

11.4.1 VIRUS STOCKS

This is another problem area for those unfamiliar with the baculovirus system. Many workers have difficulty in obtaining high-titre stocks of recombinant AcMNPV. While this is not serious in the initial stages of recombinant virus characterization, it can present difficulties when wishing to scale-up protein production. Larger volumes of virus will be required to inoculate cells at a high multiplicity of infection; this will necessitate frequent propagation of fresh inoculum stocks.

We have found that the best method to obtain high-titre stocks of recombinant virus ($>10^8$ pfu/ml) is to use spinner cultures of insect cells in suspension. If these cell are inoculated in the mid-logarithmic phase of growth (5×10^5 cells/ml) with 0.1 pfu/cell of the recombinant virus and cultured for about 5 days, the titre of the resulting stock should be at least 10^8 pfu/ml. Failure to achieve this titre is almost certainly a consequence of using cells which are either too low in density, have overgrown or are not healthy in some other way. If your cell density is correct, then check the titre of the virus inoculum used to infect the culture. If you use too high a moi (more than 0.2 pfu/cell), you will not achieve the multiple rounds of infection that are necessary to obtain high titres of virus; the virus will simply infect all the cells and only go through one replication cycle, thereby giving a low-titre inoculum. It should also be noted that the quality of the plaque-assay will also affect the result when titrating the virus stocks. If small plaques are produced in a titration then some of them may be ignored which will lower the apparent titre of the virus.

Monolayer cultures of insect cells can produce high-titred stocks of virus, but the results are more unpredictable.

11.4.2 INFECTIOUS VIRUS DNA PURIFICATION

The method used to purify AcMNPV DNA is extremely simple. The major problems arise when a low-titre stock of virus is used as the starting material. In this case there is insufficient material to give a good yield of DNA. It is worth performing a plaque-assay on the starting material to ensure subsequent success.

If a good yield of virus DNA is attained, but it performs poorly in the co-transfection experiments (see section 11.5), this may be a consequence of physical damage to the DNA throughout the purification procedure. The virus DNA is large (130 kbp) and prone to shearing by mechanical forces. Excessive pipetting and passage through narrow apertures are to be avoided. Always harvest virus DNA from a caesium chloride gradient by puncturing the bottom of the tube; do not harvest with a syringe and needle as for plasmid maxi-preps. Extensive shearing of virus DNA may be detected by running 100 ng on an agarose mini-gel. Intact, supercoiled DNA should remain associated with the well, linear DNA will run as a single band very close to intact γ DNA. Sheared DNA will appear as a smear in agarose gels.

Some methods have recommended precipitating baculovirus DNA with ethanol to concentrate the preparation. We strongly advise against doing this; the virus DNA can be very difficult to resuspend and considerable shearing may result.

11.5 CO-TRANSFECTIONS

We strongly recommend the use of lipofectin/DOTMA, or similar reagent, in place of the usual calcium phosphate precipitation method. It is possible to reduce the amount of DNA to about 50 ng per transfection; this considerably extends the life of a DNA preparation. Our only caution with this comparatively new method is that we have noted some variation between different batches of lipofectin. We recommend that each new consignment of this reagent be tested with proven virus DNA samples to ensure reproducibility.

On occasion, persistent bacterial/yeast contamination of transfections may be a problem. This may simply be a result of poor experimental technique, which can only be rectified with care and attention to detail. Alternatively, the virus DNA or plasmid DNA preparations may be contaminated. Microbial contamination of the plasmid preparation is most common in our experience and is a consequence of using old stocks of TE to resuspend the DNA pellets. To rectify, ethanol precipitate the DNA again and resuspend in freshly sterilized TE buffer.

11.6 BACULOVIRUS PLAQUE-ASSAYS

This technique, whilst quite straightforward, has been one of the major sources of difficulty for newcomers to the baculovirus system.

11.6.1 CONDITION OF THE CELLS

It is vital that Sf cells in the exponential phase of growth are used. The cells should be at least 90% viable, and adhere to the plastic culture dishes within about 2 h of dispensing. If the cells remain floating then discard the dishes and wait until fresh cells are available. High-passage number cells sometimes give poor results; retrieve fresh cells from liquid nitrogen at regular intervals, see Chapter 5. Always carefully count the cell suspension prior to use. Photographs of correctly seeded insect cells are provided in Chapter 5. We find that better plaque-assays are obtained with Sf21 cells, rather than Sf9 cells.

11.6.2 PLAQUE-ASSAY MANIPULATIONS

Take care when removing the liquid medium prior to inoculating the dish with virus dilutions. The cells are easily dislodged from the plastic which will result in damaged monolayers.

11.6.3 GENERAL PROBLEMS

In the following paragraphs we list some of the most common problems encountered with the baculovirus plaque-assay system.

(a) Cells beneath the agarose are dead!

The temperature of the agarose overlay was too high prior to adding to the cells. The low-gelling temperature agarose used will remain liquid at temperatures of about 33°C. Alternatively, FCS may have been omitted from the overlay medium.

(b) No plaques visible

The virus titre is too low to register in the range of dilutions used in the plaque-assay. Repeat the plaque-assay with lower dilutions of virus.

(c) Plaques too small and polyhedra difficult to see within individual cells under the microscope

The cells were seeded too thickly and have reached confluence too early. This inhibits virus replication and produces the small-plaque phenotype. In extreme cases the plaques may not be visible at all.

(d) Plaques large but ill-defined

The cells were seeded too thinly and do not provide sufficient numbers for adequate staining with neutral red or X-gal.

(e) Plaques are concentrated around the outer edge of cell culture dish

The virus inoculum must be added to the centre of the cell monolayer because adsorption is a very rapid process.

(f) Holes in the cell monolayer

The cell monolayer was damaged at some stage of the plaque-assay manipulations. Take care when removing medium, and inoculating plates with virus and adding overlay medium.

(g) Cracks in the agarose overlay

The virus inoculum was incompletely removed after the 1 h adsorption period. Residual liquid on the cells prior to addition of the agarose-based medium interferes with the solidification process and produces a sector on the dish where the overlay fails to cover the cells adequately.

(h) Poor staining with X-gal for viruses with the *lacZ* gene (e.g. AcMNPV.*lacZ*)

On occasion we have found that some batches of X-gal are of poor quality and do not give good staining of the β-galactosidase enzyme. The X-gal may be adequate for M13 phage staining but insect cells appear to be more fastidious.

(i) Poor staining with neutral red

Old neutral red may precipitate on storage. Ensure that the 0.5 % (w/v) stock is fresh or has been stored in the dark.

11.7 SCREENING FOR RECOMBINANT VIRUSES

This has to be the single, major difficulty encountered by users of the system. The identification of polyhedrin-negative plaques has caused endless frustration to inexperienced and experienced workers alike. Fortunately, it is the one problem that has now been largely solved.

The first solution is to use one of the vectors produced by Vialard *et al.* (1989) which have the β-galactosidase coding sequences inserted, in addition to the foreign gene coding sequences, to facilitate screening for blue plaques. This works well but does not increase the number of recombinant plaques; consequently it may still require multiple rounds of plaque purification to obtain a clean virus stock. An additional problem is that these vectors tend to be very large and problems may be encountered inserting the foreign gene.

The method described by Kitts *et al.* (1990) is currently the favoured approach to the rapid derivation of recombinant viruses. The linearization of AcMNPV DNA prior to co-transfection greatly enhances the proportion of recombinant viruses derived. When AcMNPV.*lac*Z (Possee and Howard, 1987) is used, the virus DNA may be linearized within the *lac*Z coding sequences. This results in white, recombinant plaques against a background of blue, parental plaques. If clearly defined white plaques are harvested then over 80% of these may contain the foreign coding sequences (R.D. Possee, unpublished observations). This method also removes the necessity to produce the 'perfect' plaques which were so vital to screening for polyhedrin-negative phenotypes. Plaques may be of relatively poor quality but still produce easily identifiable blue/white phenotypes. The number of rounds of plaque purification is also reduced, to two, thereby saving more time.

11.8 INSTABILITY OF RECOMBINANT VIRUSES

This is not considered a problem at present. Large amounts of foreign DNA have been stably integrated into the AcMNPV genome without deletion (Urakawa *et al.*, 1988). The authors would be grateful to hear of any problems encountered in this area; 'failures' infrequently reach the scientific press!

11.9 POOR YIELDS OF RECOMBINANT PROTEIN

This is very difficult to answer satisfactorily, assuming that there is nothing wrong with the construction of the recombinant virus. Current views as to the reasons for poor expression vary considerably. In Chapter 3 we presented a summary of the types of protein which have been produced using the baculovirus system. In general, we can say that membrane-bound glycoproteins, secreted proteins and similar gene products are expressed relatively poorly. There are some exceptions, of course, but it does appear that, not unreasonably, the infected insect cells cannot process such gene products in large amounts. Proteins that remain in the cytoplasm or that are targeted to the nucleus are usually, but not always, expressed to high levels.

If the yields of a recombinant protein are not as expected then be sure to check the genome structure of the recombinant virus using Southern hybridization techniques and/or DNA sequencing across the cloning junctions. It is also worth investigating the levels of transcription of the recombinant mRNA; a point mutation in the polyhedrin promoter may seriously reduce steady-state RNA levels.

Ensure that you are harvesting the infected cells at the optimum time. Some proteins may not be stable in virus-infected cells. If these proteins are harvested too late, considerable amounts may be lost. Always perform an experiment to determine the optimum time for harvesting recombinant proteins; do not rely on data published by others.

Appendix A
A list of selected suppliers

American Type Culture Collection (ATCC)
12301 Parklawn Drive
Rockville
Maryland 20852
USA
Tel: (301) 881 2600

Amersham International PLC
UK and Export Sales Office
Lincoln Place
Green End
Aylesbury AP20 2TP
Tel: (0800) 515313

Ashwood Timber and Plastics Ltd
Phyfa House
Leven Road
London E14 0LN
Tel: (071) 987 6461

BDH Ltd (see Merck)

Bio-Rad Ltd
Bio-Rad House
Maylands Avenue
Hemel Hempstead HP2 7TD
Tel: (0800) 181134

Boehringer Mannheim (UK) Ltd
Bell Lane
Lewes

East Sussex BN7 1LG
Tel: (0273) 480444

Difco Laboratories Ltd
PO Box 14B
Central Avenue
East Moseley
Surrey KT8 0SE
Tel: (081) 979 9951

Flowgen Instruments Ltd
Broad Oak Enterprise Village
Broad Road
Sittingbourne
Kent ME9 8AQ
Tel: (0795) 429737

FMC BioProducts Ltd
(also see Flowgen Instruments Ltd)
5 Maple Street
Roackland
Maine 04841-2994
USA
Tel: (207) 594 3200

Gibco-BRL
Life Technologies Ltd
Unit 4, Cowley Mill Trading Estate
Longbridge Way
Uxbridge UB8 2YG
Tel: (0895) 36355

Globepharm Ltd
PO Box 89C
Esher
Surrey KT10 9ND
Tel: (0372) 65507

ICN-Flow Ltd
Woodcock Hill
Harefield Road
Rickmansworth
Herts WD3 1PQ
Tel: (0923) 774666

J. R. Scientific Inc.
(see also Sera-Lab Ltd)
PO Box 1937
Woodland
California 95695, USA

Merck Ltd
Broom Road
Poole
Dorset BH12 4NN
Tel: (0202) 745520

Natural Environment Research Council
Institute of Virology and Environmental Microbiology
Mansfield Road
Oxford OX1 3SR
Tel: (0865) 512361

Northumbria Biologicals Ltd (NBL)
South Nelson Industrial Estate
Cramlington
Northumberland NE23 9HL
Tel: (0670) 732992

Oxford Polytechnic
School of Biological and Molecular Sciences
Gipsy Lane
Headington
Oxford OX3 0BP
Tel: (0865) 819240

Pharmacia-LKB Biotechnology
Pharmacia House
351 Midsummer Boulevard
Central Milton Keynes MK9 3HP
Tel: (0908) 661101

Promega Ltd
Epsilon House
Enterprise Road
Chilworth Research Centre
Southampton SO1 7NS
Tel: (0800) 398994

Sera-Lab Ltd
Crawley Down
Sussex RH10 4FF
Tel: (0342) 716366

SGI (UK) Ltd
Avis Way
Newhaven
East Sussex BN9 0SX
Tel: (0273) 517999

Sigma Chemical Company Ltd
Fancy Road
Poole
Dorset BH17 7NH
Tel: (0800) 373731

Stratagene Ltd (see NBL Ltd)

References

Anicetti, V. R., Keyt, B. A. and Hancock, W. S. (1989) *Trends in Biotechnology*, **7**, 342–9.

Arif, B. (1986) In *The Molecular Biology of Baculoviruses* (eds W. Doerfler and P. Bohm), Springer, Berlin, pp. 31–50.

Atkinson, A., Weitzman, M. D., Obosi, L., Beadle, D. J. and King, L. A. (1990a) *Pesticide Science*, **28**, 215–24.

Atkinson, A. E., Earley, F. G. P., Beadle, D. J. and King, L. A. (1990b) *European Journal of Biochemistry*, **192**, 451–8.

Bailey, M. J., McLeod, D. A., Kang, C.-L. and Bishop, D. H. L. (1989) *Virology*, **169**, 323–31.

Barber, G. N., Clegg, J. C. S. and Lloyd, G. (1990) *Journal of General Virology*, **71**, 19–28.

Baum, J. A., Geever, R. and Giles, N. H. (1987) *Molecular and Cellular Biology*, **7**, 1256–66.

Beames, B. and Summers, M. D. (1988) *Virology*, **162**, 206–20.

Beames, B. and Summers, M. D. (1989) *Virology*, **168**, 344–53.

Berger, S. L. and Kimmel, A.R. (eds) (1987) *Guide to Molecular Cloning Techniques. Methods in Enzymology Volume 152*, Academic Press, New York and London.

Bergh, M. L. E., Naranjo, C., Mentzer, A. F., Barsomian, G. D., Bartlett, C., Hirani, S. and Rasmussen, J. R. (1990) In *Glycobiology* (eds J. K. Welply and E. Jaworski), Wiley-Liss, New York, pp. 159–72.

Berndt, N. and Cohen, P. T. W. (1990) *European Journal of Biochemistry*, **190**, 291–7.

Bishop, D. H. L and Possee, R. D. (1990) *Advances in Gene Technology*, **1**, 55–72.

Blissard, G. W. and Rohrmann, G. F. (1990) *Annual Review of Entomology*, **35**, 127–55.

Bollag, R. J., Waldman, A. S. and Liskay, R. M. (1989) *Annual Review of Genetics*, **23**, 199–225.

Brown, C. S., Van Lent, J. W. M., Vlak, J. M. and Spaan, W. J. M.

(1991) *Journal of Virology*, **65**, 2702–6.
Brown, M. and Faulkner, P. (1977) *Journal of General Virology*, **36**, 361–4.
Bustos, M. M., Luckow, V. A., Griffing, L. R., Summers, M. D. and Hall, T. C. (1988) *Plant Molecular Biology*, **10**, 475–88.
Butters, T. D. and Hughes, R. C. (1981) *Biochimica et Biophysica Acta*, **640**, 655–71.
Butters, T. D., Hughes, R. C. and Visher, R. C. (1981) *Biochimica et Biophysica Acta*, **640**, 672–86.
Cameron, I. R. and Possee, R. D. (1989) *Virus Research*, **12**, 183–200.
Cameron, I. R., Possee, R. D. and Bishop, D. H. L. (1989) *Trends in Biotechnology*, **7**, 66–70.
Carbonell, L. F., Hodge, M. R., Tomalski, M. D. and Miller, L. K. (1988) *Gene*, **3**, 409–18.
Carson, D. D., Guarino, L. A. and Summers, M. D. (1988) *Virology*, **162**, 444–51.
Chang, L. M. S., Rafter, E., Rusquet, V. R., Peterson, R. C., White, S. T. and Bollum, F. J. (1988) *Journal of Biological Chemistry*, **263**, 12 509–13.
Chisholm, G. E. and Henner, D. J. (1988) *Journal of Virology*, **62**, 3193–200.
Cochran, M. A. and Faulkner, P. (1983) *Journal of Virology*, **45**, 961–70.
Couch, J. A. (1974) *Journal of Invertebrate Pathology*, **24**, 311–31.
Davidson, D. J., Fraser, M. J. and Castellino, F. J. (1990) *Biochemistry*, **29**, 5584–90.
Delchambre, M., Gheysen, D., Thines, D., Thiriart, C., Jacobs, E., Verdin, E., Horth, A., Burny, A. and Bex, F. (1989) *EMBO Journal*, **8**, 2653–60.
Desprès, P., Girard, M. and Bouloy, M. (1991) *Journal of General Virology*, **72**, 1331–42.
Devlin, J. J., Devlin, P. E., Clark, R., O'Rourke, E. C., Levenson, C. and Mark, D. F. (1989) *Bio/Technology*, **7**, 286–92.
Dodson, M. S., Crute, J. J., Bruckner, R. C. and Lehman, I. R. (1989) *Journal of Biological Chemistry*, **264**, 20 835–8.
Domingo, D. L. and Trowbridge, I. S. (1988) *Journal of Biological Chemistry*, **263**, 13 386–92.
Ellis, L., Levitan, A., Cobb, M. H. and Ramos, P. (1988) *Journal of Virology*, **62**, 1634–9.
Emery, V. C. and Bishop, D. H. L. (1987) *Protein Engineering*, **1**, 359–66.
Entwistle, P. F and Evans, H. F. (1985) In *Comprehensive Insect Physiology, Biochemistry and Pharmacology* (eds G. A. Kerkut and L. I. Gilbert), Pergamon Press, Oxford, pp. 347–412.
Estes, M. K., Crawford, S. E., Penaranda, M. E., Petrie, B. L., Burns,

J. W., Chan, W.-K., Ericson, B., Smith, G. E. and Summers, M. D. (1987) *Journal of Virology*, **61**, 1488–94.

Fitzpatrick, P. F., Chlumsky, L. J., Daubner, S. C. and O'Malley, K. L. (1989) *Journal of Biological Chemistry*, **265**, 2042–7.

Forstová, J., Krauzewicz, N. and Griffin, B. E. (1989) *Nucleic Acids Research*, **17**, 1427–43.

Fraser, M. J. (1986) *Journal of Ultrastructure and Molecular Structure Research*, **95**, 189–95.

Fraser, M. J. (1989) In Vitro *Cellular and Developmental Biology*, **25**, 225–35.

French, T. J. and Roy, P. (1990) *Journal of Virology*, **64**, 1530–6.

French, T. J., Inumaru, S. and Roy, P. (1989) *Journal of Virology*, **63**, 3270–8.

French, T. J., Marshall, J. J. A. and Roy, P. (1990) *Journal of Virology*, **64**, 5695–700.

Fuchs, L. Y., Woods, M. S. and Weaver, R. F. (1983) *Journal of Virology*, **48**, 641–6.

Fung, M.-C., Chiu, K. Y. M., Weber, T., Chang, T.-W. and Chang, N. T. (1988) *Journal of Virological Methods*, **19**, 33–42.

Gearing, K. L. and Possee, R. D. (1990) *Journal of General Virology*, **71**, 251–62.

George, S. T., Arbabian, M. A., Rucho, A. E., Kiely, J. and Malbon, C. C. (1989) *Biochemical and Biophysical Research Communications*, **163**, 1265–9.

Giese, N., May-Siroff, M., Larochelle, W. J., Van Wyke Coelingh, K. and Aaronson, S. A. (1989) *Journal of Virology*, **63**, 3080–6.

Gordon, J. D. and Carstens, E. B. (1984) *Virology*, **138**, 69–81.

Gorman, C. (1982) In *DNA Cloning, Vol 2.* (ed. D. Glover), IRL Press, Oxford, pp. 143–90.

Grabowski, G. A., White, W. R. and Grace, M. E. (1989) *Enzyme (Basel)*, **41**, 131–42.

Grace, T. D. C. (1962) *Nature*, **195**, 788–9.

Graham, F. L. and van der Eb, A. J. (1973) *Virology*, **52**, 456–67.

Granados, R. R. and Williams, K. A. (1986) In *The Biology of Baculoviruses, I. Biological Properties and Molecular Biology* (eds R. R. Granados and B. A. Federici), CRC Press, Boca Raton, Florida, pp. 89–108.

Greenfield, C., Patel, G., Clark, S., Jones, N. and Waterfield, M. D. (1988) *EMBO Journal*, **7**, 139–46.

Groebe, D. R., Chung, A. E. and Ho, C. (1990) *Nucleic Acids Research*, **18**, 4033.

Grula, M. A., Buller, P. L. and Weaver, R. (1981) *Journal of Virology*, **38**, 916–21.

Grunstein, M. and Hogness, D. (1975) *Proceedings of the National*

Academy for Sciences, USA, **72**, 3961–4.
Guarino, L. A. and Summers, M. D. (1986a) *Journal of Virology*, **57**, 563–71.
Guarino, L. A. and Summers, M. D. (1986b) *Journal of Virology*, **60**, 215–23.
Guarino, L. A., Gonzalez, M. A. and Summers, M. D. (1986) *Journal of Virology*, **60**, 224–9.
Hammock, B. D., Bonning, B. C., Possee, R. D., Hanzlik, T. N. and Maeda, S. (1990) *Nature*, **344**, 458–61.
Harlow, E. and Lane, D. (1988) *Antibodies, a laboratory manual*, Cold Spring Harbor Laboratory Press, New York.
Harmon, S. A., Johnston, J. M., Ziegelhoffer, T., Richards, O. C., Summers, D. F. and Ehrenfeld, E. (1988) *Virus Research*, **10**, 273–80.
Hasemann, C. A. and Capra, J. D. (1990) *Proceedings of the National Academy of Sciences, USA*, **87**, 3942–6.
Hauser, C., Fusswinkel, H., Li, J., Oellig, C., Kunze, R., Mueller-Neumann, M., Heinlein, M., Starlinger, P. and Doerfler, W. (1988) *Molecular and General Genetics*, **214**, 373–8.
Heimpel, A. M. and Adams, J. R. (1966) *Journal of Invertebrate Pathology*, **8**, 340–6.
Heinderyckx, M., Jacobs, P. and Bollen, A. (1989) *Molecular Biology Reports*, **13**, 225–32.
Herrera, R., Lebwohl, D., de Herreros, A. G., Kallen, R. G. and Rosen, O. M. (1988) *Journal of Biological Chemistry*, **263**, 5560–8.
Hilditch, C. M., Rogers, L. J. and Bishop, D. H. L. (1990) *Journal of General Virology*, **71**, 2755–9.
Hill-Perkins, M. S. and Possee, R. D. (1990) *Journal of General Virology*, **71**, 971–6.
Hink, W. F. (1970) *Nature*, **226**, 466–7.
Howard, S. C., Ayres, M. D. and Possee, R. D. (1986) *Virus Research*, **5**, 109–19.
Hu, S.-L., Kosowski, S. G. and Schaaf, K. F. (1987) *Journal of Virology*, **61**, 3617–20.
Huber, J. (1986) In *The Biology of Baculoviruses, II. Practical applications for insect control* (eds R. R. Granados and B. A. Federici), CRC Press, Boca Raton, Florida, pp. 181–202.
Iacono-Conners, L. C., Schmaljohn, C. S. and Dalrymple, J. M. (1990) *Infection and Immunity*, **58**, 366–72.
Innis, M. A., Gelfand, D. H., Sninsky, J. J. and White, T. J. (eds) (1990) *PCR Protocols, a Guide to Methods and Applications*, Academic Press, New York and London.
Inumura, S. and Roy, P. (1987) *Virology*, **157**, 472–9.
Inumaru, S., Ghiasi, H. and Roy, P. (1987) *Journal of General Virology*, **68**, 1627–35.

Jarvis, D. L. and Summers, M. D. (1989) *Molecular and Cellular Biology*, **9**, 214–23.

Jarvis, D. L., Fleming, J.-O., Kovacs, G. R., Summers, M. D. and Guarino, L. A. (1990a) *Bio/Technology*, **8**, 950–5.

Jarvis, D. L., Oker-Blom, C. and Summers, M. D. (1990b) *Journal of Cellular Biochemistry*, **42**, 181–91.

Jeang, K.-T., Giam, C.-Z., Nerenberg, M. and Khoury, G. (1987a) *Journal of Virology*, **61**, 708–13.

Jeang, K.-T., Holmgren-Konig, M. and Khoury, G. (1987b) *Journal of General Virology*, **61**, 1761–4.

Johnson, P. W., Attia, J., Richardson, C. D., Roder, J. C. and Dunn, R. J. (1989) *Gene*, **77**, 287–96.

Kang, C.-Y., Bishop, D. H. L., Seo, J.-S., Matsuura, Y. and Choe, M. (1987) *Journal of General Virology*, **68**, 2607–13.

Kaplan, G., Freistadt, M. S. and Racaniello, V. R. (1990) *Journal of Virology*, **64**, 4697–702.

Kataoka, H., Troetschler, R. G., Li, J. P., Kramer, S. J., Carney, R. L. and Schooley, D. A. (1989) *Proceedings of the National Academy of Sciences, USA*, **86**, 2976–80.

Kelly, D. C. (1987) In *Molecular Basis of Virus Disease* (eds W. C. Russell and J. W. Almond), Cambridge University Press, pp. 289–302.

Kelly, D. C. and Lescott, T. (1981) *Microbiologica*, **4**, 35–47.

Kelly, D. C. and Wang, X. (1981) *Annals de Virologie*, **132E**, 247–59.

Kelly, D. C., Brown, D. A., Ayres, M. D., Allen, C. J. and Walker, I. O. (1983) *Journal of General Virology*, **64**, 399–408.

King, L. A., Mann, S. G., Lawrie, A. M. and Mulshaw, S. G. (1991a) *Virus Research*, **19**, 93–104.

King, L. A., Kaur, K., Mann, S. G., Lawrie, A. M., Steven, J. and Ogden, J. E. (1991b), *Gene*, **106**, 151–7.

Kitts, P. A., Ayres, M. D. and Possee, R. D. (1990) *Nucleic Acids Research*, **18**, 5667–72.

Kornfield, R. and Kornfield, S. (1985) *Annual Review of Biochemistry*, **54**, 631–4.

Kozak, M. (1986) *Cell*, **44**, 283–92.

Krishna, S., Blacklaws, B. A., Overton, H. A., Bishop, D. H. L. and Nash, A. A. (1989) *Journal of General Virology*, **70**, 1805–14.

Kuroda, K., Hauser, C., Rott, R., Klenk, H.-D. and Doerfler, W. (1986) *EMBO Journal*, **5**, 1359–65.

Kuroda, K., Groner, A., Frese, K., Drenckhahn, D., Hauser, C., Rott, R., Doerfler, W. and Klenk, H.-D. (1989) *Journal of Virology*, **63**, 1677–85.

Kuroda, K., Geyer, H., Geyer, R., Doerfler, W. and Klenk, H.-D. (1990) *Virology*, **174**, 418–29.

Kuzio, J., Rohel, D. Z., Curry, C. J., Krebs, A., Carstens, E. B. and Faulkner, P. (1984) *Virology*, **139**, 414–18.

Lanford, R. E. (1988) *Virology*, **167**, 72–81.

Lanford, R. E. and Notvall, L. (1990) *Virology*, **176**, 222–33.

Lanford, R. E., Luckow, V., Kennedy, R. C., Dreesman, G. R., Notvall, L. and Summers, M. D. (1989) *Journal of Virology*, **63**, 1549–57.

Lebacq-Verheyden, J. A., Kasprzyk, P. G., Raum, M. G., Van Wyke Coelingh, K., Lebacq, J. A. and Battey, J. F. (1988) *Molecular and Cellular Biology*, **8**, 3129–35.

Lipes, M. A., Napolitano, M., Jeang, K. T., Chang, N. T. and Leonard, W. J. (1988) *Proceedings of the National Academy of Sciences, USA*, **85**, 9704–8.

Livingstone, C. and Jones, I. (1989) *Nucleic Acids Research*, **17**, 2366.

Luckow, V. A. (1990) In *Recombinant DNA Technology and Applications* (eds C. Ho, A. Prokop and R. Bajpai), McGraw-Hill, New York.

Luckow, V. A. and Summers, M. D. (1988a) *Bio/Technology*, **6**, 47–55.

Luckow, V. A. and Summers, M. D. (1988b) *Virology*, **167**, 56–71.

Luckow, V. A. and Summers, M. D. (1989) *Virology*, **170**, 31–9.

Madison, L., Travis, B., Hu, S.-L. and Purchio, A. F. (1987) *Virology*, **158**, 248–50.

Maeda, S. (1989a) *Annual Review of Entomology*, **34**, 351–72.

Maeda, S. (1989b) *Biochemical and Biophysical Research Communications*, **165**, 1177–83.

Maeda, S., Kawai, T., Obinata, M., Fujiwara, H., Horiuchi, T., Saeki, Y., Sato, Y. and Furasawa, M. (1985) *Nature*, **315**, 592–4.

Maiorella, B., Inlow, D., Shauger, A. and Harano, D. (1988) *Bio/Technology*, **6**, 1406–10.

Makino, Y., Tadano, M., Anzai, T., Ma, S.-P., Yasuda, S. and Fukunaga, T. (1989) *Journal of General Virology*, **70**, 1417–25.

Mann, S. G. and King, L. A. (1989) *Journal of General Virology*, **70**, 3501–5.

Marcy, A. I., Olivo, P. D., Challberg, M. D. and Coen, D. M. (1990) *Nucleic Acids Research*, **18**, 1207–15.

Marino, M. H. (1989) *Biopharmacology*, **2**, 18–33.

Martens, J. W. M., Hone, G., Zuidema, D., Van Lent, J. W. M., Visser, B. and Vlak, J. M. (1990) *Applied and Environmental Microbiology*, **56**, 2764–70.

Martignoni, M. E. and Iwai, P. J. (1986) *USDA Forest Service PNW-195*, Washington D.C., USGPO.

Martin, B. M., Tsuji, S., Lamarea, M. E., Maysak, K., Eliason, W. and Ginna, E. I. (1988) *DNA*, **7**, 99–106.

Matsuura, Y., Possee, R. D. and Bishop, D. H. L. (1986) *Journal of General Virology*, **67**, 1515–29.

Matsuura, Y., Possee, R. D., Overton, H. A. and Bishop, D. H. L. (1987) *Journal of General Virology*, **68**, 1233–50.

Matsuura, Y., Miyamoto, M., Sato, T., Morita, C. and Yasui, K. (1989) *Virology*, **173**, 674–82.

Mazzone, H. M. (1985) In *Viral Insecticides for Biological Control* (eds K. Maramororosch and K. E. Sherman), Academic Press, London and New York, pp. 695–734.

McBride, A. A., Bolen, J. B. and Howley, P. M. (1989) *Journal of Virology*, **63**, 5076–85.

McMaster, G. K. and Carmichael, G. G. (1977) *Proceedings of the National Academy of Sciences, USA*, **74**, 4835–8.

Merryweather, A. T., Weyer, U., Harris, M. P. G., Hirst, M., Booth, T. and Possee, R. D. (1990) *Journal of General Virology*, **71**, 1535–44.

Miller, L. K. (1981) *Journal of Virology*, **39**, 973–6.

Miller, L. K. (1988) *Annual Review of Microbiology*, **42**, 177–99.

Miller, L. K., Jewell, J. E. and Browne, D. (1981) *Journal of Virology*, **40**, 275–8.

Miller, D. W., Safer, P. and Miller, L. K. (1986) In *Genetic Engineering Vol 8* (eds J. K. Setlow and A. Hollaender), Plenum Press, New York, pp. 277–98.

Miltenburger, H. G. (1980) *Development and Biological Standards*, **46**, 295–300.

Miltenburger, H. G. and David, P. (1980) *Development and Biological Standards*, **46**, 183–6.

Miyamoto, C., Smith, G. E., Farrell-Towt, J., Chizzonite, R., Summers, M. D. and Ju, G. (1985) *Molecular and Cellular Biology*, **5**, 2860–5.

Moore, N. F., King, L. A. and Possee, R. D. (1987) *Insect Science Applications*, **8**, 275–89.

Morikawa, Y., Overton, H. A., Moore, J. P., Wilkinson, A. J., Brady, R. L., Lewis, S. J. and Jones, I. M. (1990) *AIDS Research and Human Retroviruses*, **6**, 765–73.

Morikawa, S., Booth, T. F. and Bishop, D. H. L. (1991) *Virology*, **183**, 288–97.

Murhammmer, D. W. and Goochee, C. F. (1988) *Bio/Technology*, **6**, 1411–18.

Murphy, C. I., Weiner, B., Bikel, I., Piwnica, W. H., Bradley, M. K. and Livingston, D. M. (1988) *Journal of Virology*, **62**, 2951–9.

Nissen, M. S. and Friesen, P. D. (1989) *Journal of Virology*, **63**, 493–503.

Nishikawa, K., Fukuhara, N., Liprandi, F., Green, K., Kapikian, A. Z., Channock, R. M. and Gorziglia, M. (1989) *Virology*, **173**, 631–7.

Noteborn, M. H. M., de Boer, G. F., Kant, A., Koch, G., Bos, J. L., Zantema, A. and van der Eb, A. J. (1990) *Journal of General Virology*, **71**, 2641–8.

Nyunoya, H., Akagi, T., Ogura, T., Maeda, S. and Shimotohno, K. (1988) *Virology*, **167**, 538–44.

Oker-Blom, C. and Summers, M. D. (1989) *Journal of Virology*, **63**, 1256–64.

Oldfield, S., Adachi, A., Urakawa, T., Hirasawa, T. and Roy, P. (1990) *Journal of General Virology*, **71**, 2649–56.

Ollo, R. and Maniatis, T. (1987) *Proceedings of the National Academy of Science, USA*, **84**, 5700–4.

Ooi, B. G., Rankin, C. and Miller, L. K. (1989) *Journal of Molecular Biology*, **210**, 721–36.

O'Reilly, D. R. and Miller, L. K. (1988) *Journal of Virology*, **62**, 3109–19.

Orr-Weaver, T. L., Szostak, J. W. and Rothstein, R. J. (1983) *Methods in Enzymology*, **101**, 228–45.

Overton, H. A., Ihara, T. and Bishop, D. H. L. (1987) *Virology*, **157**, 338–50.

Overton, H. A., Fujii, Y., Price, I. R. and Jones, I. M. (1989) *Virology*, **170**, 107–16.

Parekh, R. B., Dwek, R. A., Edge, C. J. and Rademacher, T. W. (1989) *Trends in Biotechnology*, **7**, 117–21.

Patel, G. and Stabel, S. (1989) *Cellular Signalling*, **1**, 227.

Patel, G., Greenfield, C., Stabel, S., Waterfield, M. D., Parker, P. J. and Jones, N. C. (1988) In *Current Communication in Molecular Biology: Viral Vectors* (eds Y. Gluzman and S. H. Hughes), Cold Spring Harbor, New York, pp. 98–103.

Paul, J. I., Tavare, J., Denton, R. M. and Steiner, D. F. (1990) *Journal of Biological Chemistry*, **265**, 13074–83.

Pendergast, A. M., Clark, R., Kawasaki, E. S., McCormick, F. P. and White, O. N. (1989) *Oncogene*, **4**, 759–66.

Pennock, G. D., Shoemaker, C. and Miller, L. K. (1984) *Molecular and Cellular Biology*, **4**, 399–406.

Piatak, M. J. A., Lane, E., O'Rourke, R., Clark, R. and Houston, L. L. (1988) *ICSU Short Reports*, **8**, 62.

Piwnica-Worms, H., Williams, N. G., Cheng, S. H. and Roberts, T. M. (1990) *Journal of Virology*, **64**, 61–8.

Podgwaite, J. D. (1985) In *Viral Insecticides for Biological Control* (eds K. Maramorosch and K. E. Sherman), Academic Press, London and New York, pp. 775–97.

Possee, R. D. (1986) *Virus Research*, **5**, 43–59.

Possee, R. D. and Howard, S. C. (1987) *Nucleic Acids Research*, **15**, 10233–48.

Possee, R. D., Weyer, U. and King, L. A. (1990) In *Control of Virus Disease* (eds N. J. Dimmock, P. D. Griffiths and C. R. Madeley), Cambridge University Press, pp. 53–76.

Possee, R. D., Sun, T.-P., Howard, S. C., Ayres, M. D., Hill-Perkins, M. and Gearing, K. L. (1991) *Virology*, **185**, 229–41.

Préhaud, C., Takehara, K., Flamand, A. and Bishop, D. H. L. (1989) *Virology*, **173**, 390–9.

Préhaud, C., Harris, R. D., Koh, C.-L., Wong, J., Flamand, A. and Bishop, D. H. L. (1990) *Virology*, **178**, 486–97.

Rankin, C., Ooi, B.G. and Miller, L. K. (1988) *Gene*, **70**, 39–49.

Ray, R., Galinski, M. S. and Compans, R. W. (1989) *Virus Research*, **12**, 169–80.

Rice, W. C., Lorimer, H. E., Prives, C. and Miller, L. K. (1987) *Journal of Virology*, **61**, 1712–16.

Rohrmann, G. F. (1986) *Journal of General Virology*, **67**, 1499–513.

Rota, P. A., Black, R. A., De, B. K., Harmon, M. W. and Kendal, A. P. (1990) *Journal of General Virology*, **71**, 1535–44.

Rusche, J. R., Lynn, D. L., Robert-Guroff, M., Langlois, A. J., Lyerly, H. K., Carson, H., Krohn, K., Ranki, A., Gallo, R. C., Bolognesi, D. P., Putney, S. D. and Matthews, T. J. (1987) *Proceedings of the National Academy of Sciences, USA*, **84**, 6924–8.

Sambrook, J., Fritsch, E. F. and Maniatis, T. (1989) *Molecular Cloning, a Laboratory Manual*, Cold Spring Harbor Laboratory Press, New York.

Schmaljohn, C. S., Sugiyama, K., Schmaljohn, A. L. and Bishop, D. H. L. (1988) *Journal of General Virology*, **69**, 777–86.

Schmaljohn, C. S., Parker, M. D., Ennis, W. H., Dalrymple, J. M., Collet, M. S., Suzich, J. A. and Schmaljohn, A. L. (1989) *Virology*, **170**, 184–92.

Sissom, J. and Ellis, L. (1989) *Biochemical Journal*, **261**, 119–26.

Smith, G. E. and Summers, M. D. (1979) *Journal of Virology*, **30**, 828–38.

Smith, G. E., Summers, M. D. and Fraser, M. J. (1983a) *Molecular and Cellular Biology*, **3**, 2156–65.

Smith, G. E., Fraser, M. J. and Summers, M. D. (1983b) *Journal of Virology*, **46**, 584–93.

Smith, G. E., Ju, G., Ericson, B. L., Moschera, J., Lahm, H. W., Chizzonite, R. and Summers, M. D. (1985) *Proceedings of the National Academy of Sciences, USA*, **82**, 8404–8.

Southern, E. M. (1975) *Journal of Molecular Biology*, **98**, 503–17.

Srinivasan, G. and Thompson, E. B. (1990) *Molecular Endocrinology*, **4**, 209–16.

St Angelo, C., Smith, G. E., Summers, M. D. and Krug, R. M. (1987) *Journal of Virology*, **61**, 361–5.

Stahl, P. D. (1990) *American Journal of Respiratory Cellular and Molecular Biology*, **2**, 317–18.

Streett, D. A. and Hink, W. F. (1978) *Journal of Invertebrate Pathology*, **32**, 112–13.

Summers, M. D. (1977) *EPA Ecologocal Research Services EPA-600/3-77-130*, Washington, D.C., USGPO.
Summers, M. D. and Anderson, D. L. (1973) *Journal of Virology*, **12**, 1336–46.
Summers, M. D. and Smith, G. E. (1987) *Texas A and M University Experimental Station Bulletin No 1555.*
Tada, A., Fuse, A., Sekine, H., Simizu, B., Kondo, A. and Maeda, S. (1988) *Virus Research*, **9**, 357–67.
Takehara, K., Ireland, D. and Bishop, D. H. L. (1988) *Journal of General Virology*, **69**, 2763–77.
Thiem, S. M. and Miller, L. K. (1989) *Journal of Virology*, **63**, 2008–18.
Thomas, P. S. (1980) *Proceedings of the National Academy of Sciences, USA*, **77**, 5201–5.
Thomsen, D. R., Post, L. E. and Elhammer, A. P. (1990) *Journal of Cellular Biochemistry*, **43**, 67–79.
Tomalski, M. D., Wu, J. and Miller, L. K. (1988) *Virology*, **167**, 591–600.
Tramper, J., Williams, J. B., Joustra, D. and Vlak, J. M. (1986) *Enzyme and Microbial Technology*, **8**, 33–6.
Tramper, J., van den End, E. J., de Gooijer, C. D., Kompier, R., van Lier, F. L. J., Usmany, M. and Vlak, J. M. (1990) *Annals of the New York Academy of Sciences*, **589**, 423–30.
Tweeten, K. A., Bulla, L. A. and Consigli, R. A. (1980) *Journal of Virology*, **33**, 866–76.
Urakawa, T. and Roy, P. (1988) *Journal of Virology*, **62**, 3919–27.
Urakawa, T., Small, D. A. and Bishop, D. H. L. (1988) *Virus Research*, **11**, 303–17.
Urakawa, T., Ferguson, M., Minor, P. D., Cooper, J., Sullivan, M., Almond, J. W. and Bishop, D. H. L. (1989) *Journal of General Virology*, **70**, 1453–63.
Vail, P. V., Sutter, G., Jay, D. L. and Gough, D. (1971) *Journal of Invertebrate Pathology*, **17**, 383–8.
Van Wyke Coelingh, K. L., Murphy, B. R., Collins, P. L., Lebacq-Verheyden, A.-M. and Battey, J. F. (1987) *Virology*, **160**, 465–72.
Vaughn, J. L., Goodwin, R. H., Tompkins. G. J. and McCawley, P. (1977) In Vitro *Cellular and Developmental Biology*, **13**, 213–17.
Vialard, J., Lalumiere, M., Vernet, T., Briedis, D., Alkhatib, G., Henning, D., Levin, D. and Richardson, C. (1990) *Journal of Virology*, **64**, 37–50.
Vik, T. A., Sweet, L. J. and Erikson, R. L. (1990) *Proceedings of the National Academy of Sciences, USA*, **87**, 2685–9.
Vissavajjhala, P. and Ross, A. H. (1990) *Journal of Biological Chemistry*, **265**, 4746–52.
Vlak, J. M. and Smith, G. E. (1982) *Journal of General Virology*, **41**, 1118–21.

Vlak, J. M. and Keus, R. J. A. (1990) *Viral Vaccines*, Wiley-Liss, New York.

Vlak, J. M., Klinkenberg, F. A., Zaal, K. J. M., Usmany, M., Klinge-Roode, E. C., Geervliet, J. B. F., Roosien, J. and Van Lent, J. W. M. (1988) *Journal of General Virology*, **69**, 765–76.

Vlak, J. M., Schouten, A., Usmany, M., Belsham, G. J., Klinge-Roode, E. C., Maule, A. J., Van Lent, J. W. M. and Zuidema, D. (1990) *Virology*, **179**, 312–20.

Volkman, L. E. (1986) *Current Topics in Microbiology and Immunology*, **131**, 103–18.

Volkman, L. E. and Keddie, B. A. (1990) *Seminars in Virology*, **1**, 249–56.

Volkman, L. E., Summers, M. D. and Hsieh, C. H. (1976) *Journal of Virology*, **19**, 820–32.

Wang, X. and Kelly, D. C. (1983) *Journal of General Virology*, **64**, 2229–36.

Wathen, M. W., Brideau, R. J. and Thomson, D. R. (1989) *Journal of Infectious Diseases*, **159**, 255–64.

Watson, C. J. and Hay, R. T. (1990) *Nucleic Acids Research*, **18**, 1167–73.

Webb, N. R., Madoulet, C., Tossi, P.-F., Broussard, D. R., Sneed, L., Summers, M. D. and Nicolau, C. (1989) *Proceedings of the National Academy of Sciences, USA*, **86**, 7731–5.

Weiss, S. A. and Vaughn, J. L. (1986) In *The Biology of Baculoviruses, II. Practical application for insect control* (eds R. R. Granados and B. A. Federici), CRC Press, Boca Raton, Florida, pp. 64–7.

Weiss, S. A., Smith, G. C., Kalter, S. S. and Vaughn, J. L. (1981) *In Vitro*, **17**, 485–502.

Weiss, S. A., Gorfien, S., Fike, R., DiSorbo, D. and Jayme, D. (1990) *Proceedings of the 9th Australian Biotechnology Conference*, University of Queensland Press, pp. 220–31.

Weyer, U. and Possee, R. D. (1988) *Nucleic Acids Research*, **16**, 3635–53.

Weyer, U. and Possee, R. D. (1989) *Journal of General Virology*, **70**, 203–8.

Weyer, U. and Possee, R. D. (1991) *Journal of General Virology*, in press.

Weyer, U., Knight, S. and Possee, R. D. (1990) *Journal of General Virology*, **71**, 1525–34.

Whitefleet-Smith, J., Rosen, E., McLinden, J., Ploplis, V. A. and Fraser, M. J. (1989) *Archives of Biochemistry and Biophysics*, **271**, 390–9.

Whitford, M., Stewart, S., Kuzio, J. and Faulkner, P. (1989) *Journal of Virology*, **63**, 1393–9.

Williams, G. V., Rohel, D. Z., Kuzio, J. and Faulkner, P. (1989) *Journal of General Virology*, **70**, 187–202.
Wilson, M. E., Mainprize, T. H., Friesen, P. D. and Miller, L. K. (1987) *Virology*, **61**, 661–6.
Yoden, S., Kikuchi, T., Siddell, S. G. and Taguchi, F. (1989) *Virology*, **173**, 615–23.
Yoo, D., Parker, M. D., Song, J., Cox, G. J., Deregt, D. and Babiuk, L. A. (1991) *Virology*, **183**, 91–8.
Zuidema, D., Schouten, A., Usany, M., Maule, A. J., Belsham, G. J., Rossien, J., Klinge-Roode, E. C., van Lent, J. W. M. and Vlak, J. M. (1990) *Journal of General Virology*, **71**, 2201–9.

Index

Those page numbers in bold refer to figures, those in italics refer to tables.

www.ingramcontent.com/pod-product-compliance
Ingram Content Group UK Ltd.
Pitfield, Milton Keynes, MK11 3LW, UK
UKHW020807290926
15470UKWH00003B/1011
9780412371509